PHYSICAL GEOLOGY

Workbook

Sixth Edition

Giuseppina Kysar Mattietti
Stacey Verardo
George Mason University

Kendall Hunt
publishing company

Kendall Hunt
publishing company

www.kendallhunt.com
Send all inquiries to:
4050 Westmark Drive
Dubuque, Iowa 52004-1840

ISBN 978-1-5249-3870-3

Published in the United States of America

Contents

Introduction

We are delighted to present the sixth edition of the *Introductory Physical Geology Workbook*.

As for the previous editions, we have provided hands-on activities that complement the physical geology lectures and guide the students from understanding science to applying their knowledge. For this edition, we have modified all the exercises to encourage students to explore relevant and up-to-date aspects of physical geology.

To strengthen the connection between lecture and laboratory experience we have designed pre-lab questions that focus on necessary information needed to do carry out the exercises. Labs are designed to run after the lecture topic has been presented. We envision that the Laboratory instructors can assign the pre-lab questions as homework or for discussion before the lab, eliminating the need to carry out a review lecture before the lab and making the beginning of the lab, generally a presentation of the topic, an interactive experience.

We have included more database-related exercises in this workbook. The availability of scientific data through databases is increasing exponentially and we believe that the students need to be able to query databases, produce meaningful representations of the data and grasp the meaning and the implications of the information that retrieved this way.

We believe that students should be aware of their own learning progress through self-regulation. This is why, at the end of each exercise, there is a rubric for the skills acquired with each laboratory. The rubric prompts the students to rate their level of confidence in practicing the exercise-specific skills, and in evaluating the level of challenge as they complete the exercise. This is a valuable self-efficacy tool as cognitive research indicates that students learn better when they reflect on their learning. For the instructor, these can be used to gauge how the students relate to the material, and can detect where conceptual misconceptions and problems may be developing.

We especially wish to thank all the students, laboratory instructors, and colleagues that have constantly given use feedback and constructive comments.

Section 1

Earth Materials

Chapter 1

Identification of Common Rock-Forming Minerals

PRELAB STUDY SESSION

In preparation to this exercise, briefly answer the following questions. Refer to your textbook for the information you need. In addition to these questions, your lab instructor might give you more questions to answer.

A. What is a mineral, how does it differ from any other solid?

B. What determines the physical properties of minerals?

C. What are the main mineral groups and how are they classified?

D. How many minerals can you recall? Without looking them up, do you know what they are used for?

MINERAL PROPERTIES USED FOR IDENTIFICATION

The following properties are usually tested for common rock-forming minerals

Crystal shape (Habit)

The perfect crystal shape (habit is a quality mostly for museum pieces), keep in mind that you will likely see pieces of crystals, rather than single well-formed crystals. Aggregates (clusters) of very small crystals are very common, so at times it is hard to describe the habit. However, in most cases, you will be able to see if the specimen shape is elongated, tabular, prismatic, and so on. These observations will be useful to discern the mineral in the descriptive section of the mineral key (Tables 1.2 and 1.3)

Luster

Luster is a description of how the mineral reflects light. For this exercise, it is best to start with evaluating if the luster is metallic or nonmetallic. Be aware that a shiny mineral is not necessarily a metallic mineral. The face of a cell phone is shiny and dark, when not in use, but it is nonmetallic. Metallic minerals reflect light like coins, so they are easy to identify but they can also easily tarnish and look dull/rusty. Scratching the surface of the specimen with a nail will show the underlying metallic luster. (See Figure 1.1 for difference in luster).

Nonmetallic luster can be described as glassy, dull, earthy, or greasy. These qualifiers will be useful when trying to discern other specific properties of minerals using the mineral key.

Fracture

Fracture is the way the mineral breaks, a mineral that fractures breaks with irregular surfaces with no two pieces having the same shape or size. Quartz, for example, shows often regularly shaped crystals but breaks with conchoidal fracture, an arc-shaped break (see Figure 1.1b).

Cleavage

Minerals that have cleavage break along surfaces of weak chemical bonds that are determined by the arrangement of the atoms in the crystalline structure. A mineral that exhibits cleavage, breaks along parallel surfaces/edges/corners of varying sizes, but all are oriented in the same direction. This is where the chemical bonds of the crystal are weaker. Some cleavage is easy to identify, for instance, the one direction of cleavage in mica, the cubic cleavage of halite and galena and the rhombohedral cleavage of calcite. Often, determining the correct number and angle of cleavage can be challenging, for example, if a mineral is fragile or soft (like graphite) cleavage surfaces can be smeared. In most cases, though, it is easy to detect whether or not cleavage is there. As you tilt the crystal to the light, the surfaces of cleavage, being parallel, will reflect light simultaneously.

Feldspars present diagnostic striations on cleavage surfaces. The striations look like irregular, lines in a color either slightly lighter or darker than the surrounding mineral. Figure 1.1d shows the relationship between crystal shape and fracture and cleavage.

a. Crystals of calcite showing the three directions of cleavage defining rhombohedral prisms. The perfect cleavage is maintained in the smallest fragments.

b. Crystals of quartz showing the prismatic habit in the full crystal (top of photo) and the fracture in the fragments of the quartz crystals (bottom of photo). The fracture of the quartz is called conchoidal. The luster is glassy.

c. Close up of a crystal of galena showing metallic luster and evident cleavage in three directions at 90°. This sample has metallic luster.

d. Fragment of a very large crystal of feldspar. The surface of cleavage is easy to identify, they form parallel irregular steps. On cleavage surfaces the striations are also visible. The luster of feldspar is nonmetallic, and can be described with words like glassy, or pearly.

e. Close up of a large fragment of hornblende amphibole. The cleavage surface defines parallel surfaces with irregular edges. The second direction of cleavage forms a 120° angle with the first set of surfaces. This sample has a nonmetallic luster, the cleavage surface shine with a glassy luster.

Note: scale in cm is the same for all images.

Figure 1.1   Properties of Common Minerals

Hardness

Hardness is the resistance to abrasion or scratching. The Mohs scale ranks the minerals from the softest to the hardest, with talc as the softest (hardness of 1) and diamond as the hardest (hardness of 10) (Table 1.1). A few common items of well-known hardness are used to test minerals: human fingernails have a hardness of about 2.5, copper penny has 3.5, a standard carpenter's nail has 4.5, a glass plate 5.5, and an unglazed ceramic tile has a hardness of 6.5.

The correct use the plate (glass or ceramic) is critical to determine hardness: always hold the plate firmly on the table and drag forcefully the sample back and forth along the plate a couple of times. Observe the plate, if you can see a definite scratch, then the mineral is harder than the plate, thus harder than 5.5. When using the nails or the copper penny hold firmly the sample on the table, and use the nail or the penny to try and make a groove on the sample.

Hardness	Minerals	Testing Items
10	Diamond	
9	Corundum	
8	Topaz	
7	Quartz	
6.5		Streak plate
6	Orthoclase	
5.5		Glass plate: Masonry nail
5	Apatite	
4.5		Carpenter nail
4	Fluorite	
3.5		Copper penny
3	Calcite	
2.5		Fingernail
2	Gypsum	
1	Tale	

Table 1.1   Mohs Hardness Scale

Mineral Color

The color of common rock-forming minerals is most often the result of impurities trapped in the crystal lattice. For example, quartz (SiO_2) can be clear, white, yellow, purple, dark, and so on. A few common minerals have a diagnostic color range; examples are sulfur is yellow, olivine is apple-green, and feldspar is orange-pink potassium. Use the color of the mineral as diagnostic only after testing for luster, hardness, and cleavage.

Streak Color

This test occurs when a mineral is scratched across a ceramic tile. Because of the tile's hardness, the mineral is crushed in the contact area, leaving behind a streak of mineral powder. The color is diagnostic especially for metallic minerals. Most nonmetallic minerals have a very-light-colored powder. A black ceramic tale is sometimes used to observe the color of powdered light minerals.

Magnetism

This property is evident only in the mineral magnetite, or in agglomerates of iron oxides that contains a significant amount of magnetite.

Reactivity to Diluted Hydrochloric Acid

When diluted HCl is dropped on a carbonate, effervescence occurs which releases CO_2. This reaction is noted easily with calcite ($CaCO_3$). The mineral (and rock) dolomite ($CaMgCO_3$) reacts weakly only if it is powdered. To test for dolomite, scratch the sample and drop a diluted HCl over the scratch. That will be the only place where the sample reacts.

Specific Gravity

This quantifies how heavy a mineral is when compared to water, which has a specific gravity of 1. To measure specific gravity (S.G.), it is necessary to have some laboratory equipment, although weighing the mineral against a known volume of displaced distilled water can work as well. Alternatively, specific gravity can be roughly estimated by hefting the sample. This technique however can only provide a comparative estimate, not a precise or accurate measure of specific gravity.

Tenacity

This property describes how a mineral resists breaking. A mineral can be brittle and shatter or can be easily deformed; for example, mica is flexible.

Double Refraction

The atomic structure of a mineral refracts light that travels across a clear specimen in two rays, producing two images of the one image seen through the crystal. A clear calcite crystal is the only common mineral that shows this property.

MINERAL IDENTIFICATION EXERCISE

Learning to identify minerals is a skill that is refined with time and practice, be patient! It is also important to keep in mind that the samples in the lab are rarely whole crystals, most of the times they are pieces of crystals or aggregates of many individual crystals. However, the properties you test for remain the same, whether you are testing a perfect museum quality specimen or a small sample found in the lab. For this exercise, it is important to take the time to observe carefully the sample from different points of view, best if the observation is carried out with a hand lens.

Identification tables, also called mineral keys, are used for this exercise (Tables 1.2 and 1.3). The keys are flowcharts for the mineral properties previously described. The keys list the generalities for a mineral, which may include the range of crystal habits and color that the mineral can display. Care should be taken that the description on the key fit the observation of the mineral. Examine carefully Figure 1.1 to see how properties like luster, cleavage fracture may appear in different minerals. The Mineralogical Society of America mineral key can be found at http://www.minsocam.org/msa/collectors_corner/id/mineral_id_keyq1.htm

To identify the minerals using the keys, follow this procedure:

I.   Choose the table to use based on the luster (metallic/nonmetallic).
   a. If the mineral is clearly a metallic mineral or very evidently tarnished/oxidized, then use Table 1.2.
   b. If the mineral is nonmetallic and you can perform the hardness test with the glass plate then use Table 1.3.
II. Determine the mineral hardness, then the color and the color of streak (for metallic minerals).
III. Read the "Other Properties" section to see if the additional information matches the properties of the sample you are identifying. If it does not match, test again for hardness and observe more carefully the sample.

1. Out of the minerals you identified today, which ones most clearly exhibit cleavage? What helped you identify this property?

2. Which of the minerals you observed today had well-developed habit?

3. Which mineral were in aggregates?

UNDERSTANDING MINERAL USES

The US Geological Survey maintains an updated database with the statistics and information on the worldwide supply and demand of mineral commodities needed for the United States. Mineral commodities include the minerals as well as manufactured products that originated from earth materials. For each of the minerals you have identified go to the US commodities statistics and information website:
https://minerals.usgs.gov/minerals/pubs/commodity/

1. Report on the worksheet the mineral main uses, mining methods and environmental impact (this information is displayed on the first paragraph for each mineral commodity or in the mineral commodity summary).
2. Now examine the map at page 12 of the publication: Mineral Commodities summary (2017). The map shows the major import sources of nonfuel mineral commodities for which the United States was greater than 50% net import reliant in 2016.
https://minerals.usgs.gov/minerals/pubs/mcs/2017/mcs2017.pdf

Based on the number of commodities, which three countries in the United States relaying most heavily on?

COMMON METALLIC MINERALS			
Hardness	Streak	Other Properties	Mineral
Not scratched by steel nail or knife H > 5.5	Dark gray	brass yellow, may tarnish brown; brittle, no cleavage, cubic crystals common. S.G. = 5.0.	PYRITE ("fool's gold") FeS_2 iron sulfide
	Dark gray	dark gray to black; magnetic, no obvious cleavage. S.G. = 5.2.	MAGNETITE Fe_3O_4 iron oxide
Scratched by steel nail or knife H ≤ 5.5	Red to red-brown	silver to gray, may have tiny glittery flakes, tarnish red. S.G. = 4.9-5.3.	HEMATITE Fe_2O_3 iron oxide
	Dark gray	golden yellow, may tarnish purple; brittle, no cleavage. S.G. = 4.1-4.3.	CHALCOPYRITE $CuFeS_2$ copper-iron sulfide
	White to yellow-brown	brown to yellow, or black; submetallic, complex cleavage. S.G. = 3.9-4.0.	SPHALERITE ZnS Zinc sulfide
Scratched by penny H ≤ 3.5	Copper	copper to dark brown, may oxidize green; malleable. S.G. = 8.8-8.9.	NATIVE COPPER Cu copper
	Gray to dark gray	silvery gray, tarnishes dull gray; cubic cleavage. S.G. = 7.4-7.6.	GALENA PbS lead sulfide
Scratched by fingernail H ≤ 3.5	Dark gray	gray to black, marks paper easily; greasy feel. S.G. = 2.1-2.3.	GRAPHITE C carbon

Table 1.2   Metallic Minerals (S.G. = *specific gravity)

COMMON NON-METALLIC MINERALS			
Hardness	Streak	Other Properties	Mineral
Scratches glass plate Not scratched by steel nail or knife H > 5.5	White	gray, red, brown, blue; greasy luster, no cleavage. S.G. = 3.9-4.1.	CORUNDUM Al_2O_3 aluminum oxide
	White	colorless, yellow, blue, or brown; one perfect cleavage, crystal faces often striated. S.G. = 3.5-3.6.	TOPAZ $Al_2O_4(OH,F)_2$ hydrous fluoro-aluminum silicate
	White	green, yellow, pink, blue, brown, or black slender crystals, no cleavage. S.G. = 3.0-3.2.	TOURMALINE complex silicate
	White	any color to colorless, transparent to translucent, greasy luster, no cleavage, conchoidal fracture. S.G. = 2.7.	QUARTZ SiO_2 silicon dioxide
	White	green or black conchoidal fracture, no cleavage, S.G. = 3.3-3.4.	OLIVINE $(Fe,Mg)_2SiO_4$ ferromagnesian sillicate
	White	dark red, brown, pink, green, or yellow; transparent to translucent, no cleavage. S.G. = 3.4-4.3.	GARNET complex silicate
	White	blue greenish gray, black, white, light grey striations on some cleavage planes; two cleavages at nearly 90°, S.G. = 2.6-2.8.	PLAGIOCLASE FELDSPAR $NaAlSi_3O_8$ to $CaAl_2Si_2O_8$ calcium-sodium aluminum silicate
	White	Pale, pink, brown, salmon, red, orange stuations are present and subparallel, two cleavagesat 90°. A blue green variety of this feldspars is called amaroute S.G. = 2.6.	POTASSIUM FELDSPAR $KAlSi_3O_8$ potassium aluminum silicate

Table 1.3 Non-Metallic Minerals

COMMON NON-METALLIC MINERALS			
Hardness	Streak	Other Properties	Mineral
Scratched by steel nail, knife H ≤ 5.5	White	dark green to black, dull, two cleavage directions that intersect at about 90° S.G. = 3.2-3.5.	PYROXENE (AUGITE) calcium ferromagnesian silicate
	Pale green	dark gray to black, opaque, two cleavage directions at 60° and 120°, slender crystals. S.G. = 3.0-3.3.	AMPHIBOLE (HORNBLENDE) calcium ferromagnesian silicate
	White	green, yellow, gray, or variegated; dull masses or asbestos fibrous crystals, no cleavage, S.G. = 2.2-2.6.	SERPENTINE $Mg_6Si_4O_{10}(OH_8)$ hydrous magnesian silcate
	Red to red-brown	red to brown, opaque, earthy luster, S.G. = 4.9-5.3.	HEMATITE Fe_2O_3 iron oxide
	Yellow-brown	yellow-brown to dark brown, amorphous, S.G. = 3.6-4.0.	LIMONITE Fe_2O_3 x nH_2O hydrous iron oxide
	White	colorless purple, blue, yellow, or green; cleavage, in 4 complex directions S.G. = 3.0-3.3.	FLUORITE CaF_2 calcium flouride
	Light blue	vivid royal blue, earthy masses or tiny crystals, effervesces in dilute HCl. S.G. = 3.7-3.8.	AZURITE $Cu_3(CO_3)_2(OH)_2$ hydrous copper carbonate
	Green	green to gray-green laminated crusts or masses of tiny, granular crystals; effervesces in dilute HCl. S.G. = 3.9-4.0.	MALACHITE $Cu_2CO_3(OH)_2$ hydrous copper carbonate
	White	white, gray, pink to brown; opaque; rhombohedral cleavage; effervesces in dilute HCl only if powdered. S.G. = 2.8-2.9.	DOLOMITE $CaMg(CO_3)$ magnesian calcium carbonate

Table 1.3 Continued

NON-METALLIC MINERALS			
Hardness	Streak	Other Properties	Mineral
Scratched by penny $H \leq 3.5$	White	to varied colors including colorless, white, yellow, gray, green,brown, red, blue; transparent to translucent, rhombohedral cleavage; effervesces in dilute HCl. S.G. = 2.7.	CALCITE $CaCO_3$ calcium carbonate
	Gray-brown	very dark brown to black, one perfect cleavage; flexible, very thin sheets. S.G. = 2.7-3.1.	BIOTITE MICA ferromagnesian potassium, hydrous aluminum silcate
	White	colorless, white, yellow, red, blue, brown; cubic crystals and cubic cleavage, salty taste. S.G. = 2.1-2.6.	HALITE NaCl sodium chloride
Scratched by fingernail $H \leq 2.5$	Pale Yellow	yellow to red, bright crystals or earthy masses, brittle, no cleavage, conchoidal fracture. Sulfur smell S.G. = 2.1.	SULFUR S sulfur
	White	colorless, yellow, brown, red-brown; one perfect cleavage; flexible, elastic sheets. S.G. = 2.7-3.0.	MUSCOVITE MICA potassium hydrous aluminum silicate
	White	Aggregate of micro crystals dark green, gray one perfect cleavage, but visible. S.G. = 2.6-3.0.	CHLORITE ferromagnesian aluminum silicate
	White	May show one good cleavage, may be fibrous, Colorless to white; easily scratched with fingernail. S.G. = 2.0-2.4.	GYPSUM $CaSO_4 \times 2H_2O$ calcium sulfate
	White	white to very light brown, earthy, comes off earily in a powder. S.G. = 2.6.	KAOLINITE $Al_4(Si_4O_{10})(OH)_8$ hydrous aluminum silicate
	White	white, gray, greenish, pink, yellow. soapy feel, pearly to greasy luster, massive or foliated, cleavage in one direction is hard to see S.G. = 2.7-2.8.	TALC $Mg_3Si_4O_{10}(OH)_2$ hydrous magnesium silicate

Table 1.3 Continued

Student Name _____ Lab section _____ Date _____

Sample	Hardness	Luster	Color/Streak	Fracture/Cleavage	Other Diagnostic Properties (reactivity to HCL, magnetism, etc.)	Mineral Name	Mineral Main uses	Mining Extraction Environmental Impact
1								
2								
3								
4								
5								
6								

Sample	Hardness	Luster	Color/Streak	Fracture/ Cleavage	Other Diagnostic Properties (reactivity to HCL, magnetism, etc.)	Mineral Name	Mineral Main uses	Mining Extraction Environmental Impact
7								
8								
9								
10								
11								
12								

Sample	Hardness	Luster	Color/Streak	Fracture/ Cleavage	Other Diagnostic Properties (reactivity to HCL, magnetism, etc.)	Mineral Name	Mineral Main uses	Mining Extraction Environmental Impact
13								
14								
15								
16								
17								
18								

Laboratory Experience Assessment: Minerals

Learning Objective	Level of Confidence		
	Hesitant, concept is unclear, would not know how to use/apply	I have a general idea of what this is about and with guidance I could apply what I learned to problem solving	I am confident I understand this topic and I can apply it to solving a problem
Test for and describe the main physical properties of minerals: luster, hardness and cleavage			
Perform the reactivity test with HCl			
Use of key for mineral identification			
How confident are you with identifying the major rock forming minerals?			
What challenged you the most about this activity? Why?			

Chapter 2

Identification of Igneous Rocks

PRELAB STUDY SESSION

In preparation to this exercise, briefly answer the following questions. Refer to your textbook for the information you need. In addition to these questions, your lab instructor might give you more questions to answer.

A. Which minerals are found in igneous rocks?

B. How do cooling conditions affect texture and composition of igneous rocks?

C. How does the presence of volatiles affect the texture and composition of igneous rocks?

D. Intrusive rocks form only well below the surface, so how is that we can find intrusive rocks on top of mountains?

E. What does the geothermal gradient tell us about melting rocks at increasing depth?

CLASSIFICATION OF IGNEOUS ROCKS

Igneous rocks are classified according to two criteria: texture and composition. The texture depends mostly on the cooling history of the magma. The main types of igneous rocks texture are briefly described in the following section and are illustrated in Figure 2.1

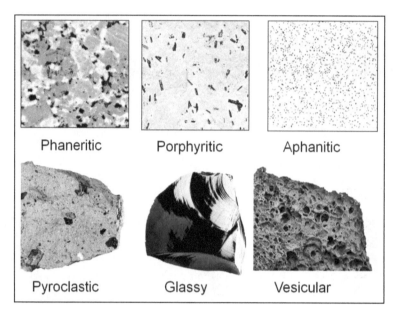

Figure 2.1 Example of average igneous rocks textures. Field of view is 4 × 4 cm for each image.

TEXTURE

Glassy

This texture forms when lava cools so rapidly that crystals cannot develop. Rocks with this texture have a distinctive conchoidal fracture, identical to that of quartz. This is the texture of volcanic glass; the color may vary to include shades of red and brown colors from oxides present in the cooling glass.

Aphanitic

The crystals are very small and in general cannot be discerned individually with the naked eye. Some may be visible when using a hand lense. This texture feels slightly gritty at the touch. Many volcanic (extrusive) rocks have this texture.

Phaneritic

The crystals are large enough to be seen with the naked eye. This texture is typical of plutonic (intrusive) rocks that formed from slow cooling of magma. Some phaneritic rocks may have a distinctively bimodal distribution of crystalize, this happens especially when the magma moves to shallower depths while crystallizing. This texture is called **Porphyry**.

Pegmatitic

Pegmatites are phaneritic rocks characterized by average crystal sizes of 2.5 cm and above. Like the phaneritic rocks, pegmatites crystallize in a plutonic environment, however their crystallization is different from that of the other plutonic bodies, so care should be taken before defining a pegmatite based exclusively on a hand specimen.

Porphyritic

This texture is characterized by a very evident bimodal distribution of crystal sizes. The coarse size crystals are called phenocrysts and are formed by slow cooling at depth in the magma chamber. They are easy to see with the naked eye. The smaller crystals formed after eruption, in a fast cooling mode, constitute the rock groundmass (matrix). This texture is the most common texture of all volcanic rocks. Porphyritic rocks can vary significantly in the relative size and amount phenocrysts and groundmass, but they always maintain the bimodal distribution of crystal sizes.

Vesicular

This texture describes volcanic rocks with a significant amount of holes left from escaping volatiles (H_2O and CO_2) degassing lava on the surface of lava flows (scoria) or from tephra forming during explosive volcanic eruptions (e.g., pumice, lapilli).

Pyroclastic

This texture is characterized by angular fragments of various sizes and shapes. There are significant variations within this texture, so it is often necessary to obtain information about the site where these rocks were collected. Pyroclastic rocks are the product of explosive volcanic eruptions.

COMPOSITION

Igneous rock composition is based on the percentage content of Silica as highlighted in the Bowen Crystallization series. Compositionally igneous rocks are grouped in felsic (silica rich), intermediate, and mafic (silica poor, but rich in Fe Mg bearing silicates). Ultramafic rocks bear mostly olivine and pyroxene. When using hand samples, the following three mineralogy criteria are used to classify igneous rocks compositionally.

Quartz

Quartz is a major component of felsic rocks, it is rare or uncommon in intermediate and generally absent in mafic rocks.

Feldspars

Orthoclase (potassium feldspar) is very common in felsic rocks. In mafic and intermediate rocks only low sodium members of the plagioclase group are present.

Mafic minerals

These minerals are accessory in felsic rocks, are more abundant in intermediate and predominant in mafic rocks. Mafic minerals in felsic rocks are mostly biotite, in intermediate mostly amphiboles and in felsic rocks mostly pyroxenes and olivine in ultramafic rocks.

Classification Criteria

The classification of igneous rocks is based on a combination of both textural and compositional criteria arranged in a table that displays composition on the horizontal axis, and texture on the vertical axis. The identification is aided visually by the overall color of the rock: light dominant colors (white, gray, and all shades of yellow-orange-red) indicate felsic compositions, whereas dark colored rocks (mostly dark grey to black and all shades of green) indicate a mafic composition; intermediate composition rocks never contain less than 30% of either mafic or felsic minerals (Table 2.1).

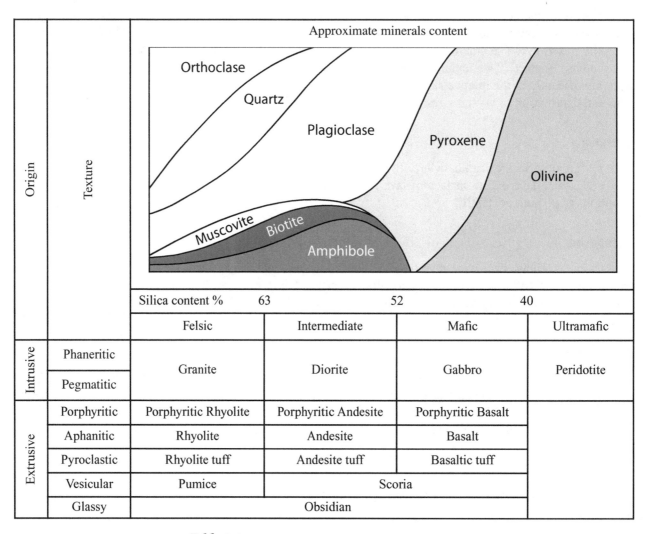

Origin	Texture	Felsic	Intermediate	Mafic	Ultramafic
		Silica content % 63	52	40	
Intrusive	Phaneritic	Granite	Diorite	Gabbro	Peridotite
	Pegmatitic				
Extrusive	Porphyritic	Porphyritic Rhyolite	Porphyritic Andesite	Porphyritic Basalt	
	Aphanitic	Rhyolite	Andesite	Basalt	
	Pyroclastic	Rhyolite tuff	Andesite tuff	Basaltic tuff	
	Vesicular	Pumice	Scoria		
	Glassy	Obsidian			

Table 2.1   Chart for Igneous Rocks Identification

EXERCISE PROCEDURE

Follow the steps below to classify the samples, and use the worksheet to record your observations:

1. Determine the texture.
2. Estimate the percentage of light versus dark minerals using the chart in Figure 2.2. If you have an aphanitic rock then estimate the overall color tone as light, dark, or intermediate. If the rock is vesicular, do not consider the holes in this estimate.
3. Estimate the relative percentage of quartz and feldspars. Quartz looks grey, plagioclase appears white or light grey in felsic rocks, and with the hand lens it is possible to see the cleavage. Potassium feldspar is generally salmon-reddish in color.
4. Note that the most reliable classification is done when it is possible to determine the mineralogy accurately. With aphanitic rocks, you will have to rely on the overall color. For example, you will be able to determine the percentage of potassium feldspar in granite but not in a rhyolite, so you will have to rely on the overall color as percentage of felsic minerals present in the sample.
5. Use the classification chart provided in Table 2.1 to find the name of the rock.
6. Finally, connect your identification to cooling history, the possible igneous environment of formation (plutonic, volcanic), and the eruption process (lava flow, presence of volatiles, explosive eruption, etc.).
7. To aid in the estimate of the percentage use the scale in Figure 2.2

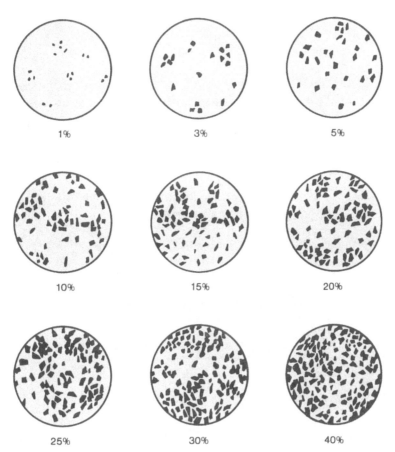

Figure 2.2 Scale for use when estimating volume percent abundance of minerals in igneous rocks. Percent abundance of dark mineral is listed below each example. From *Physical Geology: Laboratory Manual, 6th Edition* by R. D. Dallmeyer. Copyright © by Kendall Hunt Publishing Co. Reprinted by permission.

Student Name _____ Lab section _____ Date _____

Igneous Rocks Identification Worksheet

Sample	Texture	Composition (Felsic, mafic, or intermediate)	Percentage of Silicate Present by Volume	Rock Name	Cooling History
1					
2					
3					
4					
5					
6					

Igneous Rocks Identification Worksheet

Sample	Texture	Composition (Felsic, mafic, or intermediate)	Percentage of Silicate Present by Volume	Rock Name	Cooling History
7					
8					
9					
10					
11					
12					

REFLECT ON THIS EXERCISE: ANSWER THESE SYNTHESIS QUESTIONS

1. Identify the silicates visible in the phaneritic rocks you examined today and write a general sentence to describe how the minerals in the table below look like in the rocks you identified for this exercise.

Mineral	Description of the Visible Crystals (include size and any property that you can test for)
Quartz	
Plagioclase	
Potassium Feldspar	
Hornblende	
Pyroxene	
Biotite	
Muscovite	

2. Which of the rocks you examined today has the most complex cooling history? How can you tell from the observations you made?

3. Sketch the system volcano-magma chamber and represent diverse types of igneous rocks intrusive and extrusive. Give a sense of the scale.

Laboratory Experience Assessment: Igneous Rocks

Learning Objective	Level of Confidence		
	Hesitant, concept is unclear, would not know how to use/apply	I have a general idea of what this is about and with guidance I could apply what I learned to problem solving	I am confident I understand this topic and I can apply it to solving a problem
Recognize and describe the various texture of igneous rocks in hand sample			
Distinguish between intrusive and extrusive rocks			
Recognize felsic minerals in the phaneritic samples			
Recognize mafic minerals in the phaneritic samples			
Estimate the percentage of mafic and felsic minerals			
Use of key for igneous rocks identification			
How confident are you with identifying common igneous rocks?			
What challenged you the most about this activity? Why?			

Chapter 3

Identification of Sedimentary Rocks

PRELAB STUDY SESSION

In preparation to this exercise, briefly answer the following questions. Refer to your textbook for the information you need. In addition to these questions, your lab instructor might give you more questions to answer.

A. Name one clastic sedimentary rock, and explain the processes that created it.

B. Name one chemical/biochemical sedimentary rock, and explain the processes that created it.

C. What is the difference between a high-energy and low-energy environment? What rocks types illustrate these environments?

D. Explain what a well-rounded, well-sorted clastic rock represents.

CLASSIFICATION CRITERIA FOR SEDIMENTARY ROCKS

The wide range of composition and textures common in sedimentary rocks may be separated into clastic and chemical/biochemical categories (Tables 3.1 and 3.2). This rock group may also contain fossils.

Clastic sedimentary rocks are classified first on grain size (i.e., course vs. fine), and then subdivided according to composition (i.e., quartz vs. rock fragments).

Biochemical and chemical rocks (nonclastic) are classified on the basis of composition and textural characteristics such as fossil content, hardness, and particle type,

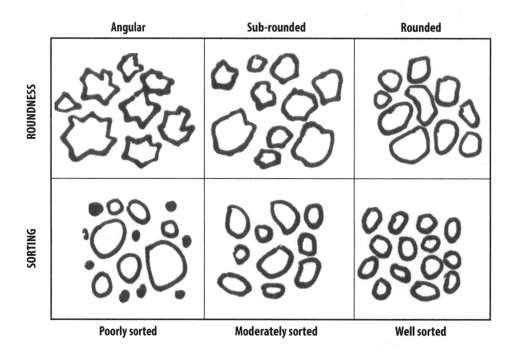

Figure 3.1 Texture characteristics of clastic rocks: roundness and sorting

TEXTURES OF CLASTIC SEDIMENTARY ROCKS

Mineral grains in clastic rocks are described in terms of sorting and roundness (Figure 3.1).

- **Sorting** describes size relationships. A rock that is poorly sorted contains various grain sizes, whereas a well-sorted rock contains grains nearly equivalent in size.
- **Roundness** describes the smoothness of the individual grains. A well-rounded grain is one with smooth edges (not necessarily spherical!), while an angular grain is one with sharp edges.

Sorting and roundness are used to determine the depositional history. In general, rocks that are well-sorted and well-rounded indicate more weathering and transportation (and thus more elapsed time) before lithification

TEXTURES OF CHEMICAL/BIOCHEMICAL ROCKS

The terms massive, crystalline, amorphous, and fossiliferous can all be used to describe chemical and biochemical rocks.

- Crystalline texture is characteristic of sedimentary rocks composed of interlocking crystals (i.e., rock salt).
- Amorphous (or cryptocrystalline) texture describes dense rocks composed of fine, noncrystalline material often deposited by chemical precipitation (i.e., chert).
- Fossiliferous textures are composed of any type of limestone ($CaCO_3$) that contain fossils or trace fossils.

EXERCISE PROCEDURE

1. Examine the rocks and determine whether the sample is clastic or chemical/biochemical (nonclastic). Most chemical/biochemical rocks are soft and can be scratched with a nail. One notable exception is chert, which is composed of silica (H=7) and will scratch glass
2. If the rock is clastic, determine the grain size and rock name using Table 3.1. For sandstone, determine the abundance of quartz, feldspar, and rock fragments for further subdivision. Breccia and conglomerate are subdivided on the amount of rounding.
3. If the rock is nonclastic, you must determine the composition of the primary material using Table 3.2.

Table 3.1 Clastic (Detrital) Sedimentary Rocks

	Grain Size	Characteristics	Rock Name	
Grains are Visible	Coarse Grained (> 2 mm) *Gravel*	Well-rounded pebbles or gravel; generally very poorly sorted	CONGLOMERATE	
		Very angular fragments of pebbles or gravel; generally poorly sorted	BRECCIA	
	Medium Grained (2 mm–¹⁄₁₆ mm) *Sand*	Sand-sized quartz grains; generally well-rounded, well-sorted; may be iron-stained	Sandstones	QUARTZ SANDSTONE
		Sand-sized, angular, feldspar grains mixed with quartz grains; generally poorly sorted; commonly pink to light red color		ARKOSE (Quartz-Feldspar Sandstone)
		Sand-sized quartz grains and rock fragments mixed with clay; common "salt and pepper" look		GRAYWACKE
Grains are Not Visible	Fine Grained (¹⁄₁₆ – ¹⁄₂₅₆ mm) *Silt*	Clay with fine quartz grains; slightly "gritty" texture	Mudstone	SILTSTONE
	Very Fine Grained (< ¹⁄₂₅₆ mm) *Clay*	*Laminated* — Smooth-feeling mixture of clay		SHALE
		Non-Laminated		CLAYSTONE

Table 3.2 Depositional Environments of Clastic Sedimentary Rocks

Environment		Sedimentary Rocks and Characteristics
Continental	Desert, Dunes	Sandstone—*well sorted, large crossbeds*
	River Beds	Sandstone, Conglomerate—*ripple marks, crossbeds*
	Floodplain	Siltstone, Shale—*mudcracks, ripple marks, lamination*
	Alluvial Fans	Conglomerate, Arkose, Sandstone—*poorly sorted, crossbeds*
	Lakes	Shale, Siltstone—*fine ripple marks, lamination*
	Glaciers	Breccia, Sandstone—*poorly sorted, angular to rounded particles*
Transitional	Delta	Sandstone, Siltstone, Shale—*crossbeds, ripple marks*
	Beach	Sandstone—*fine to medium grained, well-sorted; crossbeds*
	Tidal Flat	Mudstone, Siltstone, Sandstone—*ripple marks, crossbeds, mudcracks, lamination*
Marine	Shelf/Platform	Sandstone, Shale—*crossbeds, ripple marks*
	Slope/Rise	Mudstone, Graywacke—*graded bedding*

Table 3.3 Non Clastic Rocks, Biochemical and Chemical Sedimentary Rocks

Composition		Characteristics		Rock Name
Calcium Carbonate (CaCO₃)	EFFERVESCS WITH HC1	Cemented shell fragments	LIMESTONES	COQUINA
		Abundant fossils in massive calcareous matrix		FOSSILIFEROUS LIMESTONE
		Aggregate of small, well-rounded spherical ooids		OOLITIC LIMESTONE
		Massive and soft; composed of calcareous microfossils		CHALK
		Massive, microcrystalline lime mudstone; conchoidal fracture; dark gray to black		MICRITE
		Shows irregular laminations and voids		TRAVERTINE
		Only effervesces when ground into a powder		DOLOSTONE
Plant Material		Brown plant fibers; poorly consolidated; very lightweight		PEAT
		Low density, block, matte luster, harder than peat		LIGNITE (Low-Grade Coal)
Halite (NaCl)		Rock salt; crystalline with salty taste	EVAPORITES	ROCK SALT
Gypsum (Hydrous CaSO₄)		Massive to crystalline; can be scratched with fingernail		ALABASTER
SiO₂		Cryptocrystalline; dense; hard; has conchoidal fracture		CHERT

Table 3.4 Depositional Environments of Biochemical and Chemical Rocks

Environment		Sedimentary Rocks
Continental	Swamp	Lignite, Peat
	Caves, Hot Springs	Travertine (limestone)
	Arid, hot, restricted circulation	Evaporites (alabaster, rock salt)
Transitional	Beach	Coquina
	Tidal Flat	Evaporites
Marine	Shelf/Platform	Various limestones
	Reef	Fossiliferous limestone
	Deep Marine	Chert, Chalk, Micrite

SEDIMENTARY ROCK IDENTIFICATION WORK SHEET

Sample Number	Composition Mineralogy %	Clastic (Size, Sorting, Roundness) OR Biochemical–Chemical (HCl reactivity, Organic matter, Crystalline precipitates)	Rock Name	Depositional Environments
1				
2				
3				
4				
5				
6				

Student Name _____ Lab section _____ Date _____

SEDIMENTARY ROCK IDENTIFICATION WORK SHEET

Sample Number	Composition Mineralogy %	Clastic (Size, Sorting, Roundness) OR Biochemical–Chemical (HCl reactivity, Organic matter, Crystalline precipitates)	Rock Name	Depositional Environments
7				
8				
9				
10				
11				
12				

SYNTHESIS

1. Using Figure 3.2 below, and Tables 3.2 and 3.3, name a sedimentary rock that would likely form as a result of lithification of the sediments in each depositional environment labeled.

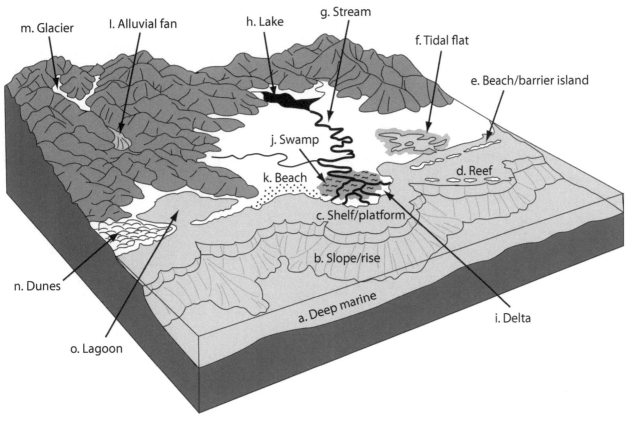

Figure 3.2 Environments of deposition.

a –   i –

b –   j –

c –   k –

d-   l –

e –   m –

f –   n –

g –   o –

h –

2. Which of the samples in your set formed the farthest from the source area? Give evidence.

3. What textural and mineralogical properties can be used to infer the distance between the depositional environment in which the rock formed and the source of the fragments?

Student Name _____ Lab section _____ Date _____

Laboratory Experience Assessment: Sedimentary Rocks

Level of Confidence

Learning Objective	Hesitant, concept is unclear, would not know how to use/apply	I have a general idea of what this is about and with guidance I could apply what I learned to problem solving	I am confident I understand this topic and I can apply it to solving a problem
Determine the clastic or chemical/biochemical nature of the rock			
Recognize the grains from matrix and cement in coarser sedimentary rocks and/or the presence of visible crystals/fossils			
Estimate grain size, sorting and roundness of a sedimentary rock			
Identify the major minerals that make up the rock			
Suggest a possible sedimentary environment in which the rock may have formed			
Use of key for sedimentary rocks identification			
How confident are you with identifying common sedimentary rocks?			
What challenged you the most about this activity? Why?			

Chapter 4

Identification of Metamorphic Rocks

PRELAB STUDY SESSION

In preparation to this exercise, briefly answer the following questions. Refer to your textbook for the information you need. In addition to these questions, your lab instructor might give you more questions to answer.

A. Explain the importance of heat and pressure in creating these rocks

B. Explain the differences metamorphic environments from which form nonfoliated and foliated rocks.

C. What is an index mineral?

MINERAL COMPOSITION

Minerals present in the original rock will adjust to the new conditions of the metamorphic environment and will reorient and/or recrystallize. These rocks do not form from magmas or fragments of preexisting rocks, and thus the overall chemical composition of the parent rock (protolith) will not vary much.

Quartz, feldspars, micas, and calcite are common minerals in metamorphic rocks. Other minerals may form due to the migration of ions and molecules that occur during metamorphic conditions. Some of these minerals are garnet, chlorite, talc, and kyanite. If a mineral forms in a very specific range of temperature and pressure, it is called an index mineral. Figure 4.1 summarizes the composition, rock type, and grade of typical regional metamorphic environments.

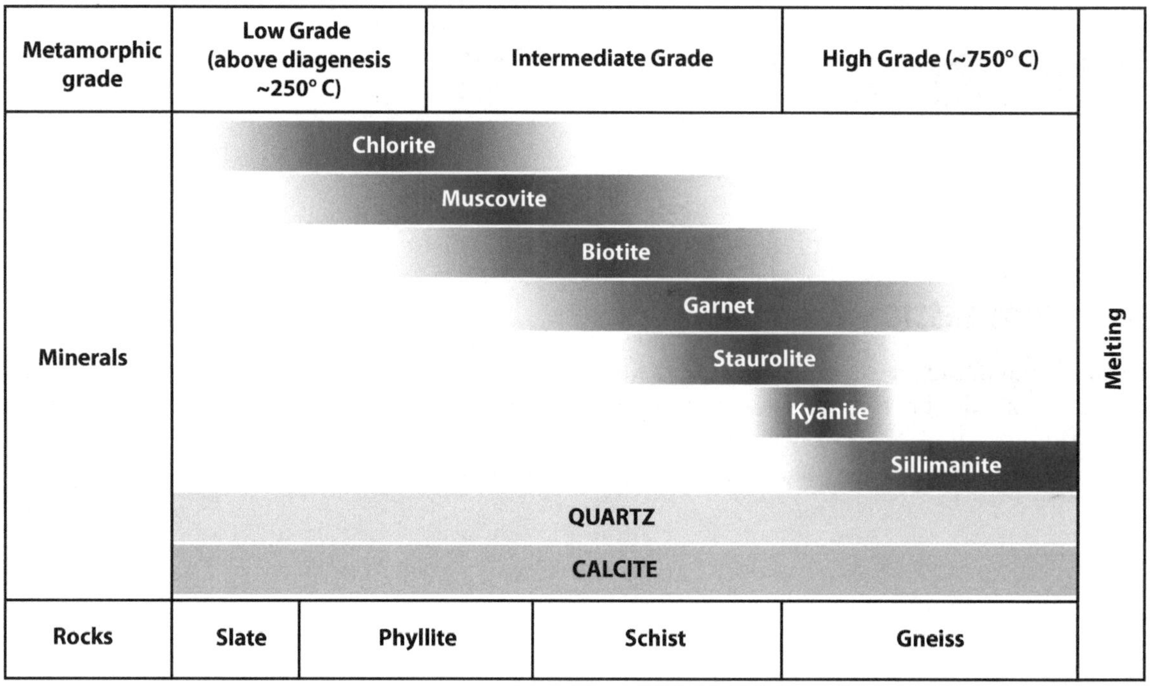

Metamorphic grade	Low Grade (above diagenesis ~250° C)	Intermediate Grade		High Grade (~750° C)	Melting
Minerals	Chlorite	Muscovite	Biotite / Garnet / Staurolite / Kyanite	Sillimanite	
	QUARTZ				
	CALCITE				
Rocks	Slate	Phyllite	Schist	Gneiss	

Figure 4.1 Minerals associated with metamorphic grade. The "range" of each mineral illustrates its stability within a certain metamorphic grade.

TEXTURE

Metamorphic rocks are described by differing textures that reflect the variation in temperature, pressure, and the presence of fluids that created them.

There are two groupings of metamorphic textures, foliated and nonfoliated. Foliated rocks contain minerals that are oriented in parallel to subparallel directions. The most common foliated textures are slaty cleavage, schistosity, and gneissic banding. These textures represent low to high metamorphic grades.

Slaty Cleavage

This foliation is generated by the parallel reorientation of small clay minerals that form perpendicular to the direction of greatest pressure. The minerals in this texture are always microscopic in size. The texture causes the rocks to "cleave" along parallel planes. It indicates a low grade of metamorphism in a regional metamorphic environment.

Schistosity

Subparallel platy minerals like, muscovite and chlorite dominate this texture. The minerals in these rocks are often macroscopic.

Gneissic banding

This texture is indicative of the highest degree of metamorphism. To form this texture, ions migrate by mass transfer and minerals produce compositionally different bands of irregular length and thickness. These rocks are predominately composed of silicate minerals.

Nonfoliated

Nonfoliated textures exhibit no orientation of its minerals and/or grains. The crystals are relatively uniform in size and mostly composed of one type of mineral. Examples include quartz, calcite, and dolomite. These textures are typical of contact metamorphism.

EXERCISE PROCEDURE

The identification of common metamorphic rocks is based on texture, followed by composition. You will use Table 4.1 to classify the metamorphic rocks.

Determine of the rock is foliated or nonfoliated. Be sure to observe the rock sample from various angles—often foliation is evident on some (not all) sides of the sample.

1. Identify the type of foliation.
2. Estimate the mineral composition and abundance.
3. Identify the metamorphic grade (Figure 4.1).
4. Identify the metamorphic environment (i.e., contact or regional) if possible.
5. Identify the parent rock (protolith).
6. Determine the name of the rock.

	Texture	Description of Texture and Composition	Rock Name	Possible Protolith	Possible Metamorphic Environment
Foliated	**Slaty cleavage**	Very dense, fine grained. Mostly made of clay minerals, which are easily scratched with a nail. Cleavage may cross the sedimentary layers.	Slate	Mudstone, shale	Regional Low grade
	Phyllitic	Dense, with a distinctive sheen luster that forms from fine-grained mica crystals invisible to the naked eye. The foliation appears on a gently wavy and/or crenulated (thinly folded) surface.	Phyllite	Slate	Regional Low to medium grade
	Schistosity	Coarsely aligned crystals of platy minerals; often micas give these rocks a shiny luster. These rocks contain index minerals such as garnet, kyanite, staurolite, etc.	Schist	Phyllite	Medium grade
	Gneissic texture	Crystalline rocks with compositional banding, resulting in irregular alternating layers of light and dark minerals, mostly silicates.	Gneiss	Igneous rocks Arkose, Greywacke, lithic sandstone Schist	High metamorphic grade
Non-foliated		Very dense, crystalline, formed by interlocking crystals of quartz of very uniform size. Unlike sandstone, rock samples are broken through rather than between quartz grains, so the rocks have a highly crystalline, almost sugary appearance. Rocks may vary in colors.	Quartzite	Sandstone	Contact metamorphism
		Dense, coarsely crystalline, composed of interlocking calcite grains. It may present "swirls" of darker oxides that were mobilized during recrystallization (veined marble).	Marble	Limestone	
		Massive jet black, low density, shiny surface with evident conchoidal structure	Anthracite	Bitiminous coal	Regional metamorphism

Table 4.1 Chart for the identification of common metamorphic rocks

Student Name _____ Lab section _____ Date _____

Metamorphic Rocks Identification Worksheet

Sample	Texture	Mineral Composition	Rock Name	Protolith	Metamorphic Grade and Environment
1					
2					
3					
4					
5					

Student Name _____ Lab section _____ Date _____

Metamorphic Rocks Identification Worksheet

Sample	Texture	Mineral Composition	Rock Name	Protolith	Metamorphic Grade and Environment
6					
7					
8					
9					
10					

REFLECT ON THIS EXERCISE: ANSWER THESE SYNTHESIS QUESTIONS

1. What effect would the pressure (indicated by the arrows) have on the minerals and the fragments illustrated in the four sketches below (Figure 4.2)? Complete the figures.

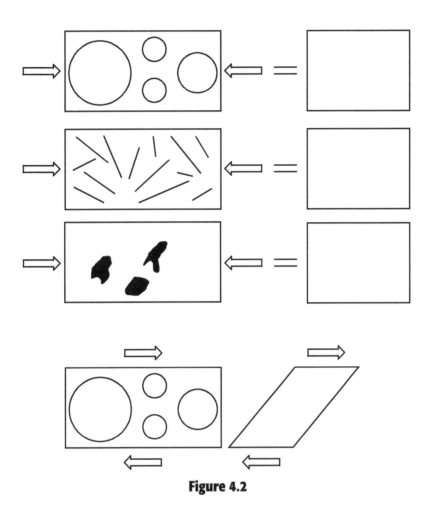

Figure 4.2

2. Rocks formed by contact metamorphism do not show foliation. Why?

3. Consider a scenario in which the gneiss in your set is brought to melting conditions. What rock will form?

Why? Explain your answer.

4. Explain the effects from erosion of a mountainside composed of gneiss. What rocks can form from weathered gneiss?

Laboratory Experience Assessment: Metamorphic Rocks

Learning Objective	Level of Confidence		
	Hesitant, concept is unclear, would not know how to use/apply	I have a general idea of what this is about and with guidance I could apply what I learned to problem solving	I am confident I understand this topic and I can apply it to solving a problem
Distinguish between foliated and non-foliated metamorphic rocks			
Recognize the main minerals that make up the metamorphic rock			
Evaluate the metamorphic degree of the rock from macroscopic evidence			
Suggest a probable protolith for a given metamorphic rock			
Use of key for metamorphic rocks identification			
How confident are you with identifying common metamorphic rocks?			
What challenged you the most about this activity? Why?			

Section 2

Surface Processes

Chapter 5

Introduction to Topographic Maps

Topographic maps are used in landform analysis to represent the size, shape, and geomorphic features of the Earth's surface. In this exercise, you will learn to read topographic maps to study various geologic processes and examine landform development.

A map is a two-dimensional representation of a portion of the Earth's surface. Maps can be represented in many projection formats. This workbook will explore topographic maps produced with the Mercator projection for which both lines of latitude and longitude are straight and parallel (Figure 5.1).

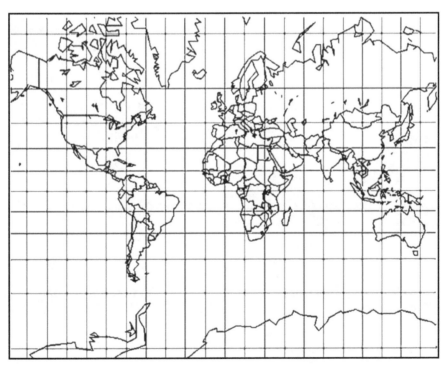

Figure 5.1 Mercator projection. Note the distortion representing the landmasses at high latitudes. This map has a scale of 1:2,000,000.

SCALE

Most maps represent a reduced image of a larger area. The amount of reduction is defined by a scale and may be expressed in the following ways:

Fractional Scale

Fractional scale is a ratio between a distance measured on a map and an equal distance measured on the Earth's surface. This ratio is termed the *representative fraction, or RF*. For example, a fractional scale of 1:24,000 indicates that one distance unit on the map (whether in inches, feet, centimeters, etc.), equals 24,000 of the *same* distance on the Earth's surface.

Graphic Scale

Graphic scale is a calibrated bar printed on the bottom of a topographic map and is used to convert distances on the map to actual ground distances. The graphic scale is divided into two parts. To the right of the zero, the scale is marked in full units of measure and is called the primary scale. To the left of the zero, the scale is divided into tenths and is called the extension scale (Figure 5.2).

Verbal Scale

Verbal scale is expressed as a specific distance on a map being equal to a specific distance on the ground. For example, one inch on a map may be stated as being equivalent to one mile on the surface.

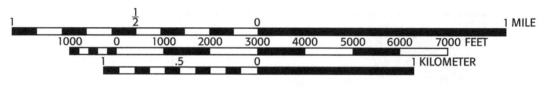

CONTOUR INTERVAL 20 FEET
NATIONAL GEODETIC VERTICAL DATUM OF 1929

Figure 5.2 Graphic scale.

LATITUDE AND LONGITUDE

Latitude and longitude are the coordinate system used to locate any point on Earth. They are given using degrees, minutes, and seconds of arcs.

Latitude

Latitude is the geographic coordinate that specifies the north-south position of a point on the Earth's surface and ranges from 0° at the equator to 90° at both the North and South Poles. **Parallels** are lines of constant latitude and circle the globe from east to west.

Longitude

Longitude is a geographic coordinate that specifies the east-west position of a point on the Earth's surface. It is an angular measurement, usually expressed in degrees. Points with the same longitude lie in lines running from the North Pole to the South Pole, these are called **meridians**. The *Prime Meridian* passes through Greenwich, United Kingdom, and is considered zero. Because of this, each point on the Earth surface can be identified by a unique latitude and longitude value set of values (Figure 5.3).

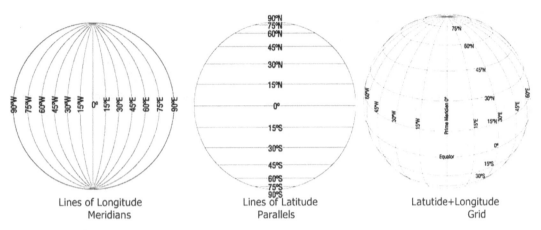

Figure 5.3 Spherical diagram showing the grid of latitude and longitude.

Contour Lines

Topographic maps illustrate the three-dimensional configuration of the Earth's surface with contour lines, continuous lines that connect points of equal elevation. They can be visualized as resulting from the intersection of a series of equally spaced, horizontal planes with the Earth's surface (Figure 5.4).

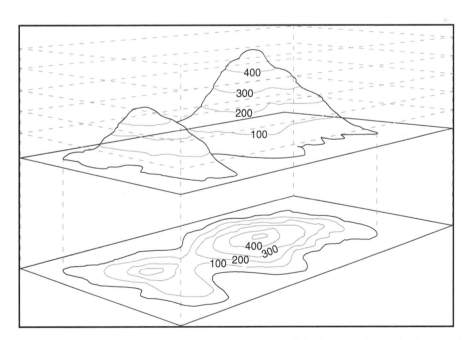

Figure 5.4 The relationship between a topographic map and the land surface: a landscape view of a mountainous island (top) and its representation in a map using contour lines (bottom).

The distance between adjacent contour lines is the **contour interval** (CI). The CI of a topographic map is in general indicated at the bottom of the map, below the scale. On each map, every 4th or 5th contour line is thicker and darker, this is an **index contour** and is labeled with its elevation (Figure 5.5).

The following rules apply when reading or creating a topographic map.

- Contours lines do not cross.
- The spacing of the contour lines is related to the steepness of the slope: closely spaced contour lines indicated steep slopes and widely spaced contour lines characterize gentle slopes.
- When contour lines go across streams they form "V"s, which point upstream.
- A series of closed concentric contours indicate a hilltop or a mountaintop
- Enclosed depressions like a sinkhole are shown by use of hachure lines on contours. Hachures point to the center of the depression.

1. Look at the map in Figure 5.5, it is a section of the Fairfax topographic map. Mark on it all the features described in this section.

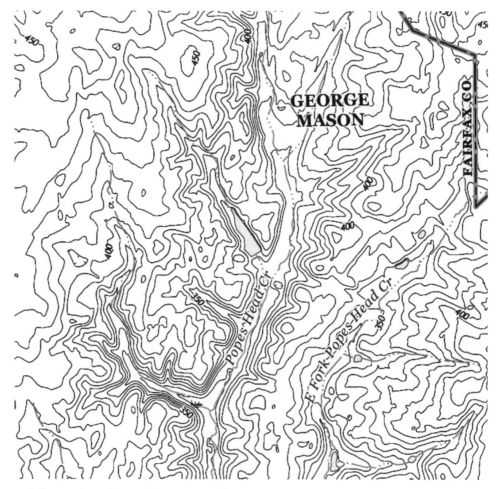

Figure 5.5   Section of the Fairfax topographic quadrangle showing contour lines distinctive features. Map's original scale is 1:24,000.

MAP ORIENTATION AND MAGNETIC DECLINATION

As a general rule, maps are oriented with north at the top. When using a compass, it is necessary to correct the magnetic measurement for the **magnetic declination**. The magnetic declination is the difference between true North (the axis around which the Earth rotates) and magnetic north (the direction the needle of a compass will point). The position of the magnetic north is not fixed, but it changes with time because the position of the magnetic poles changes. The declination is positive when the magnetic north is east of true north (Figure 5.6). Magnetic declination is usually printed on the map to the left of the scale bar at the bottom of a USGS 7.5 quadrangle.

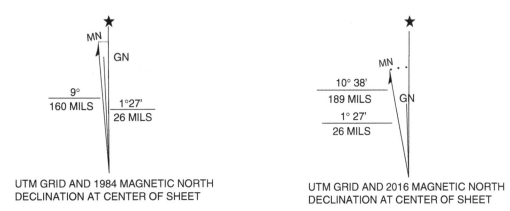

Figure 5.6 Magnetic declination. Example of the change in magnetic declination is over a 32-year period (1984–2016) for the Fairfax quadrangle topographic map. A compass aligns itself with magnetic north (MN), while GN represents geographic north. True north is represented by the star on the symbol.

MAP SYMBOLS

Standardized symbols and colors are used on maps to designate various features: water is blue, elevation and contour lines are brown, and vegetation is green. Man-made features are often in black, while roads may be in red. A booklet explaining all the symbols on USGS topographic maps (valid for most topographic maps around the world) can be found at:

https://pubs.usgs.gov/gip/TopographicMapSymbols/topomapsymbols.pdf

ELEVATION AND RELIEF

Elevation refers to the vertical distance of a point above sea level and has the same general meaning as **altitude**. The precise elevations of several points are determined with a ground survey before making the map. Surveyed locations are called bench marks, marked on topographic maps with the letters BM; they give a specific elevation said location. The elevation of a point of a topographic map is often approximated by adjacent contour lines. For example, a point between 550 and 570 would be approximately 560, while a point closer to the 570 line would be closer to 568 or 569.

The difference in elevation between two features on the map is called *local relief* (Figure 5.7). *Maximum relief* refers to the difference in elevation between the highest and lowest points within a map area.

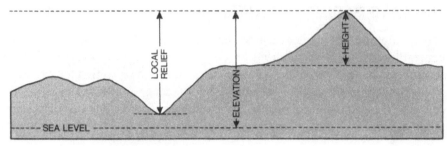

Figure 5.7 Topographic profile illustrating the differences between elevation, relief, and height.

GRADIENT

Gradient is the change in vertical height (measured in feet or meters) compared to the horizontal length of a sloping surface (measured in miles or kilometers). For example, if two points have elevations of 175′ and 125′ and are separated by 2 miles on a map, the gradient is calculated as follows:

$$\text{Gradient (ft./mile)} = \frac{\text{difference in elevation}}{\text{horizontal distance on map}} = \frac{175' - 125'}{2 \text{ mi.}}$$

$$= 50 \text{ ft./2 mi.}$$
$$= 25 \text{ ft./mi.}$$

Stream gradients are calculated with a similar procedure, BUT the map distances must be measured along the stream channels. To calculate a stream gradient, mark the spots where the stream crosses two contour lines. Measure the length of the stream between the two points where the stream crosses the contours. Because streams do not follow straight lines, you may have to measure the length of the stream in a series of smaller segments (Figure 5.8). The stream gradient is the sum of the individual segments and divided by the elevation difference of the two contour lines.

$$\text{Stream gradient} = \frac{70' - 50'}{s1 + s2 + s3 + s4 + s5 + s6}$$

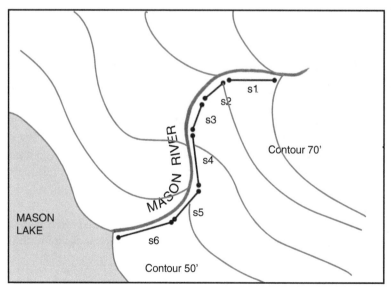

Figure 5.8 Calculating stream gradients from a topographic map. The stream crosses the 70′ and 50′ contour lines and its channels are divided into a series of nearly straight segments, s1 through s6.

TOPOGRAPHIC PROFILES

A topographic profile represents a side-view of the surface (Figure 5.8). Construction of a profile is not difficult if the following steps are followed:

1. Place a strip of paper along a selected profile line (Figures 5.9a and b).
2. Mark on the paper the exact place where each contour line, hill crest, and valley crosses the profile line and label each elevation (Figure 5.9b).
3. Place the paper above the calibrated vertical grid and project each mark at the proper elevation (Figure 5.8c).
4. Connect all points on the grid with a smooth line. The result is the profile (Figure 5.8c).

The horizontal scale of the topographic profile will be equal in length to the map from where the profile was constructed.

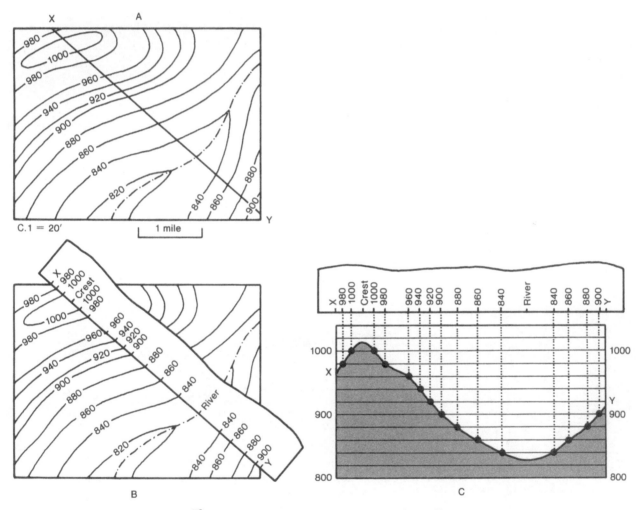

Figure 5.9 Steps to create a topographic profile.

VERTICAL EXAGGERATION

In many cases, little variation in elevation would be displayed if a vertical grid of the same scale as the horizontal grid is used. Therefore, most profiles have some degree of *vertical exaggeration* (VE) to emphasize topographic variations. Vertical exaggeration is produced by using a vertical grid with a larger scale than that on the map. The amount of vertical exaggeration should *always* be noted of the topographic profile. It may be calculated by dividing the fractional scale of the vertical grid by the fractional scale of the map. For example, if a vertical grid is constructed so 1 inch = 100 feet, it has a fractional scale of 1:1,200 (1 inch = 1,200 inches).

For a map scale 1:24,000, the vertical exaggeration is calculated as follows

$$V.E. = Vertical\ exaggeration = \frac{vertical\ grid\ map}{map\ scale}$$

$$= \frac{1:1,200}{1:24,000} = \frac{24,000}{1,200} = 20\times$$

This exaggeration indicated that vertical distances shown on the profile are magnified 20 times in comparison to actual topographic variations within the area. The effect of changing the vertical exaggeration of a topographic profile is illustrated in Figure 5.10.

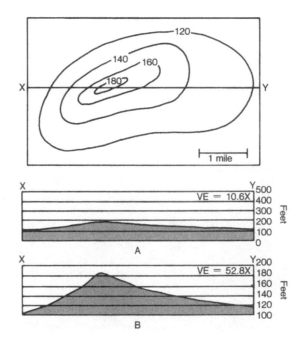

Figure 5.10 Diagram illustrating how changing vertical exaggeration affects the detail of topographic profiles. Note that increasing the vertical exaggeration from 10.6 (profile A) to 52.8 (profile B) provides a more detailed view of variations in topography.

FIND THE LATITUDE AND LONGITUDE OF A POINT ON A TOPOGRAPHIC MAP

The map used for this exercise is a topographic map from the 7.5-minute series because the latitude difference between the north and south boundary of the map is 7.5 minutes of a degree. Likewise, the difference between the maximum and minimum longitude of the map boundaries is 7.5 minutes.

To find the location of a point on this topographic map (and on any other topographic map) follow this procedure:

1. Find the length of the map boundaries: Measure map length from South to North ("A" in Figure 5.11) and measure the map width from East to West ("B" in figure 5.11). Be consistent with the units, use inches or centimeters.
2. Find the *Latitude / length* ratio: divide 7.5 minutes by the measurement "A." Do the same for the longitude (divide 7.5 by "B") and you will have the *Longitude / Length*. You will have two different values for the latitude and the longitude ratio, respectively.
3. Let's assume that you want to find latitude and longitude of a feature represented by the star in Figure 5.11. Draw a line parallel to A and call it Ay, measure Ay from the South East corner of the map. Similarly, draw another line parallel to the longitude (B), call it Bx. Measure the distance of Bx from the North–East corner.
4. Multiply Ay by *Latitude/Length* and add the result to the minimum latitude of the map to obtain the latitude of your chosen location on the map, in this case, the star. Be prepared to convert the result in minutes and seconds. Similarly, multiply Bx by *Longitude/Length* and add the result to Minimum Longitude to obtain the longitude of your chosen location on the map.

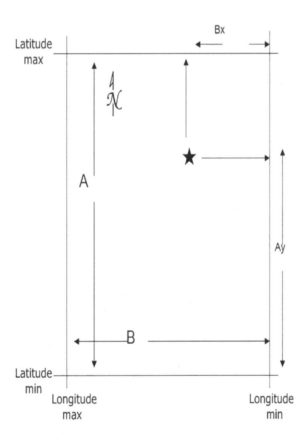

Figure 5.11 How to calculate latitude and longitude of a point on a map. The dot on the map represents Exploratory Hall; the procedure, however, is valid for any other point on the map.

CALCULATING AN AREA ON A TOPOGRAPHIC MAP

Calculating areas on topographic maps is done routinely by dedicated software; it is important, however, to understand that the calculation of any area is based on the principle to reduce any shape to a polygon formed by a collection of small regular elements scaledas needed. To calculate areas with arcGIS use the measure area tool, go to: http://www.arcgis.com/home/webmap/viewer.html?webmap=8c1408acf6d1413bbd99aa3ff149b9e2

The area of an irregularly shaped feature can be measured using a grid as demonstrated in Figure 5.12. The area is estimated by counting the number of squares that are a part of the feature to be measured. An accepted method of counting the squares is by first counting the full squares and then putting together partial squares to approximate additional full squares. (Figure 5.12b)

Alternatively, the irregular shape of a topographic feature can be approximated to a set of regular polygons area easy to calculate with (Figure 5.12c).

When calculating the area, it is necessary to convert the measurement using the maps scale. For example, an area that is 1 square cm on a 1:24,000 map corresponds to an area that is equal to 0.0576 square kilometers.

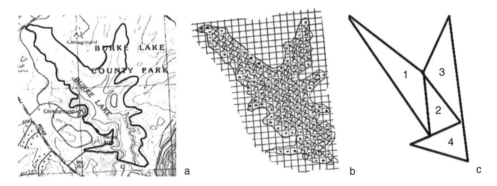

Figure 5.12   a) topographic map detail showing the outline of Burke Lake (scale 1:24,000); b) area of Burke lake estimated using a grid; c) area of the lake estimated from a set of regular polygons 1–4.

MAPPING SOFTWARE

Freeware mapping applications are available for download at no charge. Students should become aware of these resources and learn how to use them with the short tutorials available on the respective websites. Two of such apps are suitable for studying topography:

GeoMapApp

GeoMapApp is an application available free of charge on the web at: http://www.geomap app.org/index.htm

The application can be downloaded to any computer or tablet and provides direct access to the Global Multi-Resolution Topography (GMRT) compilation for bathymetry (from multibeam data) and altimetry from ASTER (Advanced Spaceborne Thermal Emission and Reflection Radiometer) and NED (National Elevation Dataset) topography datasets for the global land masscs. It is possible to find latitude, longitude, and distance among points and to carry out topographic profiles with GeoMapApp (see tutorials on GeoMapApp website).

Google Earth

To explore topography, it is useful to learn to how find geographic reference points, measure distances, and prepare elevation profiles from Google Earth, freeware software, available at https://www.google.com/earth/.

TOPOGRAPHIC MAP EXERCISE

1. Convert into ground distance the following map distance for each of the following representative fractions (RF):

    a. 1:24,000   1 inch on map = _____ feet

       1 cm on map = _____ m

       10 cm on map = _____ km

    b. 1:62,500   5 inch on map = _____ feet

       10 cm on map = _____ km

       1 inch on map = _____ feet

2. A map has an RF of 24,000. The vertical scale of a topographic profile for this map is 1 inch = 1,000 feet. What is the V.E.?

3. Calculate the average gradient for a slope that drops 200 m in elevation over 10 km. You're your calculations.

4. A topographic profile has a V.E. of 10 and a horizontal scale with an R.F. of 1:10,000. What is the vertical scale of the profile?

5. Reading contour lines:
   a. Match the contour map with the correct profile.

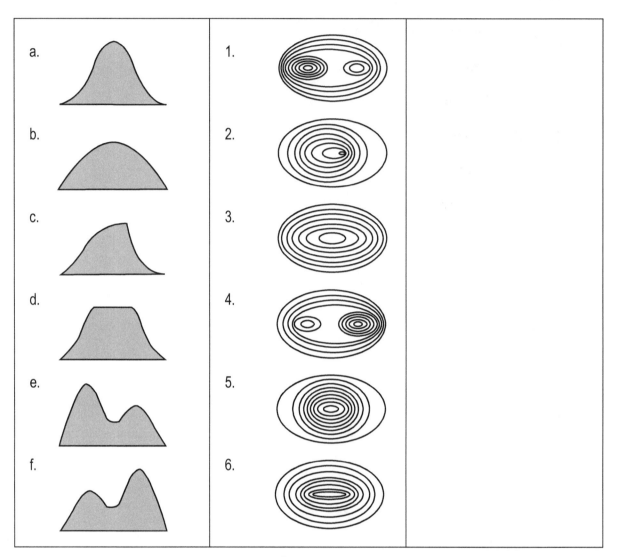

Figure 5.13

   b. Sketch a profile for contours 1 and 6 perpendicular to the profiles provided on the figure above.

6. Constructing a topographic map

To become more familiar on how contour lines work, build a topographic map using the elevation of points on the ground. In Figure 5.9, try and complete the contour lines for the map using the given points. Use a 10 CI.

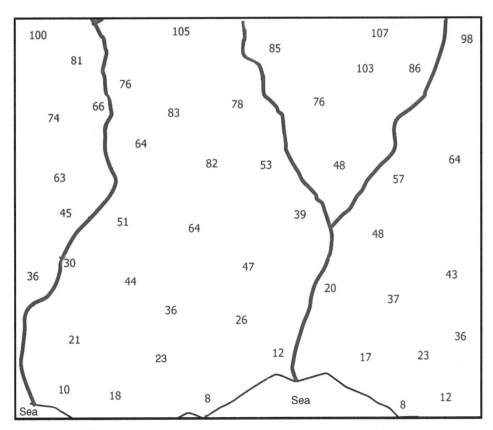

Figure 5.14   Elevation data for constructing a topographic map

7. Fairfax Topographic Map Study

   a. What is the RF of the map?

   b. What is the CI of the map?

   c. If the CI were 100 ft, would there be a greater or lesser number of contour lines?

   d. What is the maximum and minimum latitude of the map?

   e. What is the maximum and minimum longitude of the map?

   f. What is the distance (in degrees, minutes, and seconds) from the northern to the southern edges of the map?

   g. What is the distance in miles and in kilometers from the southern to northern edges of the map?

   h. What is the total area of the map? Show your calculations.

i. Choose a site on the map and find its latitude and longitude.

j. Compare with an earlier (older) version of the map. When was each published?

k. Briefly describe the changes that have occurred since the earlier map was published.

l. What is the magnetic declination of both maps? Why has it changed?

m. What is the maximum relief of the map? To accomplish this, find the highest and the lowest points of elevation. Describe where each was found on the map.

n. Identify a large water body and give the elevation of the water surface. What is the area covered by the lake? Include calculations.

o. Identify a stream on the map. In which direction does it flow?

p. How does this direction compare with the direction of the drainage pattern on the map?

8. Use the graph paper on the following two pages to construct a topographic profile.
   Horizontal scale _____

   _____

   Vertical scale _____

   _____

   Vertical Exaggeration_____

   _____

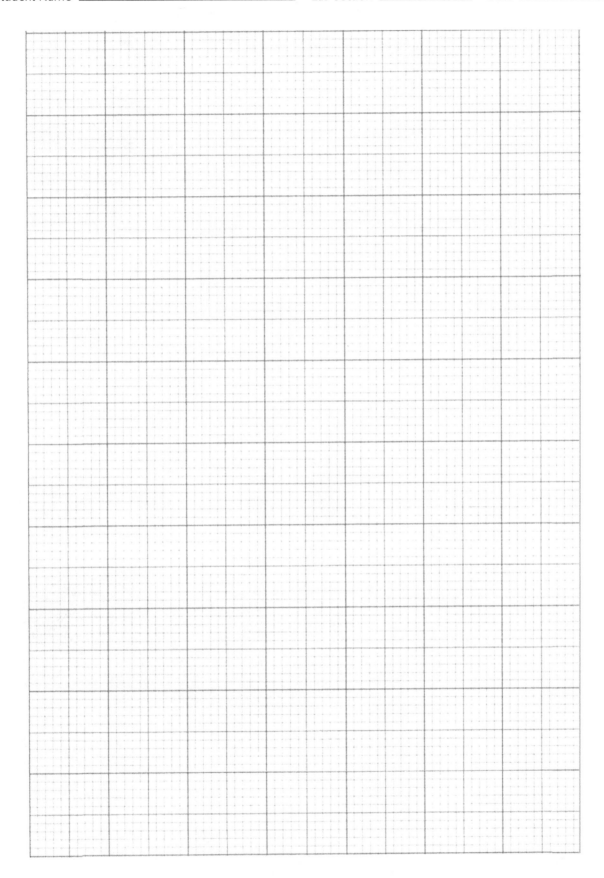

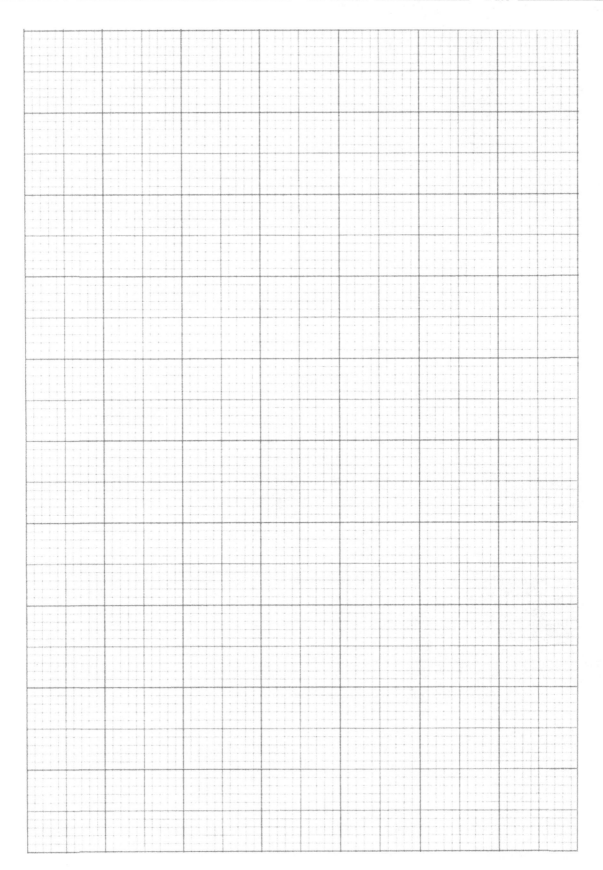

Laboratory Experience Assessment: Topographic Maps

Level of Confidence

Learning Objective	Hesitant, concept is unclear, would not know how to use/apply	I have a general idea of what this is about and with guidance I could apply what I learned to problem solving	I am confident I understand this topic and I can apply it to solving a problem
Describe the boundaries of a quadrangle map in terms of latitude and longitude (max, min)			
Locate a point on the map using latitude and longitude			
Use the scale of a map to measure distances			
Know what the various symbols on the map means			
Use the contour lines to determine elevation, relief and gradient			
Construct a topographic profile			
Determine the vertical exaggeration			
How confident are you with working with topographic maps?			
What challenged you the most about this activity? Why?			

Chapter 6

Landscape Evolution by Stream Erosion

PRELAB STUDY SESSION

In preparation to this exercise, briefly answer the following questions. Refer to your textbook for the information you need. In addition to these questions, your lab instructor might give you more questions to answer.

A. Sketch and explain the different landforms that dictate the four dominant types of drainage patterns (i.e., dendritic, trellis, radial, rectangular).

B. Sketch and label the features that you can observe in a river valley in a degradational, balanced, aggradational, and terminal stage.

C. Explain the drainage pattern around George Mason University. You may use the map from last week or Google Earth.

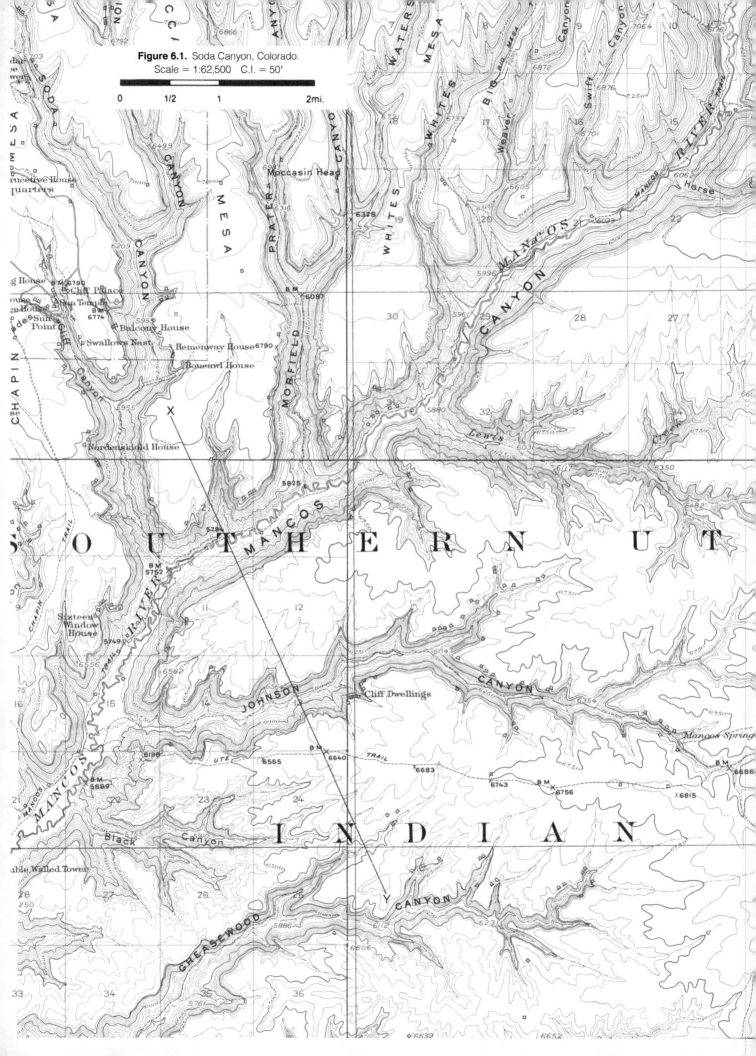

Figure 6.1. Soda Canyon, Colorado.
Scale = 1:62,500   C.I. = 50'

0   1/2   1   2mi.

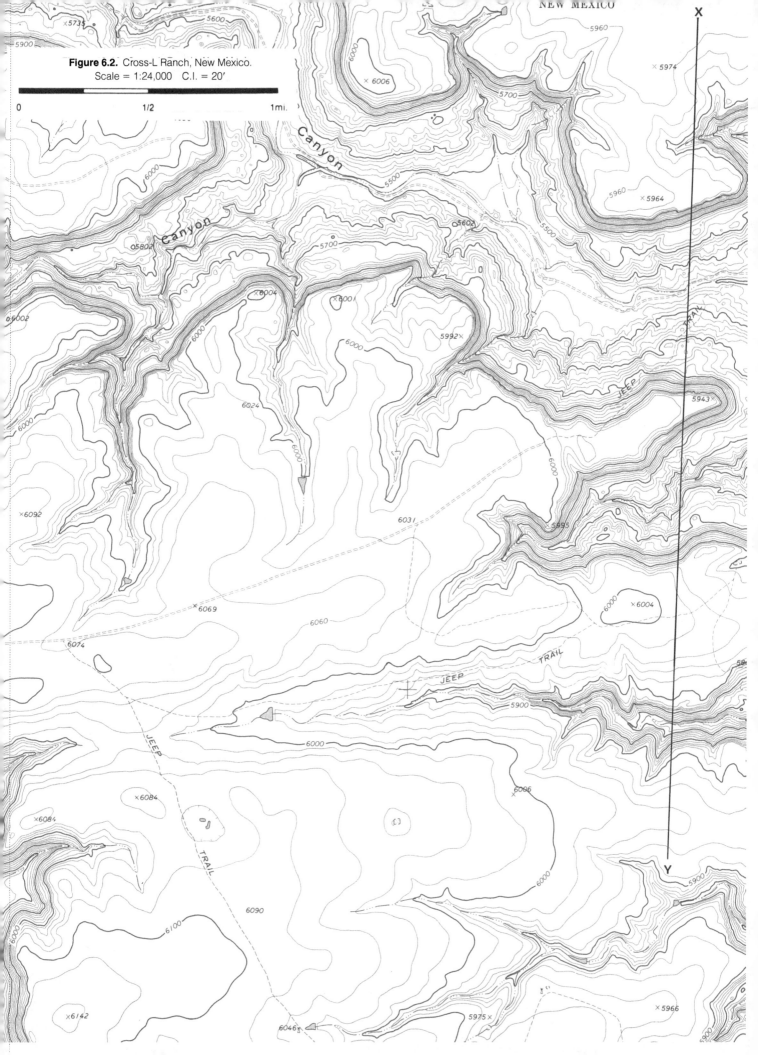

Figure 6.2. Cross-L Ranch, New Mexico.
Scale = 1:24,000   C.I. = 20′

0   1/2   1mi.

Figure 6.3. Alps Mesa, New Mexico.
Scale = 1:24,000   C.I. = 20′

0   1/2   1mi.

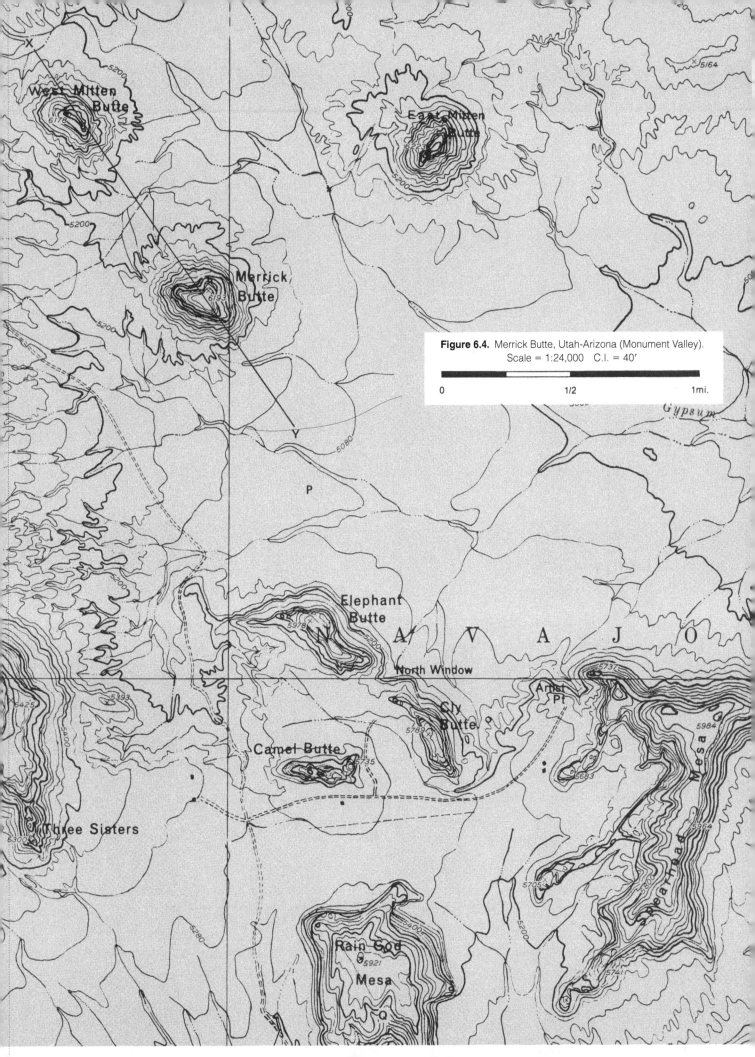

Figure 6.4. Merrick Butte, Utah-Arizona (Monument Valley).
Scale = 1:24,000   C.I. = 40′

0   1/2   1mi.

EXERCISE QUESTIONS

1. Locate the study areas for this exercise
   a. Using the marking tools in Google Earth, locate the geographic coordinates of the approximate center of the areas represented in these maps (use the marker tool in Google Earth, or calculate them following the method illustrated in Chapter 5 of this workbook).

   Soda Canyon: LAT _____ LONG _____

   Cross-L Ranch: LAT _____ LONG _____

   Alps Mesa: LAT _____ LONG _____

   Merrick Butte: LAT _____ LONG _____

   b. How far apart are these sites from each other? (estimate the distances in Km)

   c. What type of drainage do you see for each?

   Soda Canyon _____

   _____

   Cross-L Ranch _____

   _____

   Alps Mesa _____

   _____

   Merrick Butte _____

   _____

2. **SODA CANYON, COLORADO**

Refer to Figure 6.1. Look carefully at the contours of this area to visualize the landscape. Answer the following questions:

a. Describe in your own words what the landscape is like. Sketch a cross-section along the X-Y line.

b. Describe the overall shape of the upland surface.

c. Describe the cross-sectional shape of the stream valleys.

d. Measure the depth and width along the line X-Y for these valleys:

	Width	Depth
Mancos Canyon		
Johnson Canyon		
Greasewood Canyon		

e. Which canyon is older than the others? How can you tell?

3. **CROSS L RANCH, NEW MEXICO**

Refer to Figure 6.2. Look carefully at the contours of this area to visualize the landscape. Answer the following questions:

a. Draw a topographic profile along the X-Y line (use graph paper provided or GeoMappApp or Google Earth).

b. Describe the landscape for this area.

c. Describe the shape of the upland surface.

d. Descripe the shape of the stream valleys.

e. Measure the depth and width along the line X-Y for these valleys:

	Width	Depth
Northern Canyon		
North Middle Canyon		
South Middle Canyon		
Southern Canyon		

f. Which canyon is older than the others? How can you tell?

4. **ALPS MESA, NEW MEXICO**

Refer to Figure 6.3. Look carefully at the contours of this area to visualize the landscape. Answer the following questions:

a. Draw a topographic profile along the X-Y line (use graph paper provided or GeoMappApp or Google Earth).

b. Describe the landscape for this area.

c. Describe the shape of the upland surface.

d. Describe the shape of the stream valleys.

e. How does the width and depth of the valleys in this area compare to those of the previous two maps?

f. Do you think that this stream landscape is more or less developed (mature) than those on the previous two maps? Justify your answer.

5. MERRICK BUTTE, ARIZONA, MONUMENT VALLEY

Refer to Figure 6.4. Look carefully at the contours of this area to visualize the landscape. Answer the following questions:

a. Draw a topographic profile along the X-Y line (use graph paper provided or GeoMappApp or Google Earth).

b. Describe the landscape for this area.

c. Describe the shape of the upland surface. How does the areal extent of the upland compare to that of the other maps?

d. Describe the shape of the stream valleys (hint: use the valley between East Mitten Butte and Elephant Butte).

e. Measure the depth and width of a valley on the X-Y line. How does it compare to the previous maps?

f. Do you think that the stream landscape is more or less developed (mature) than those on the other three maps? Justify your answers.

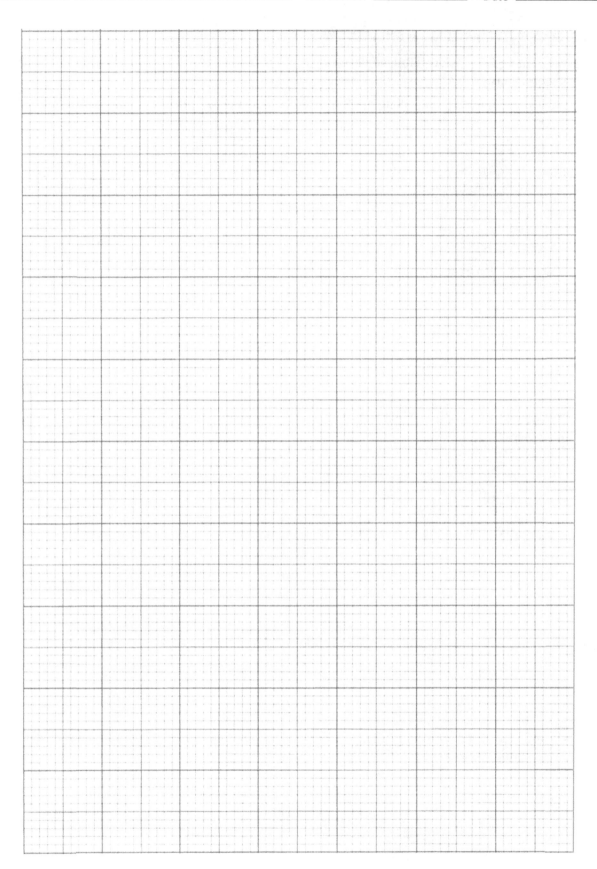

STREAM MORPHOLOGY CASE STUDIES, ON EARTH

For this exercise use Google Earth. You can "fly" to the locations in this exercise by entering the geographical coordinated given here. To complete this exercise, you need to download the basic version of Google Earth on your computer.

1. Grand Teton National Park—Oxbow Bend of the Snake River (Wyoming)
   Latitude: 43°51 59.77 N; Longitude: 110°32 56.37 W

   a. What type of river stage is this, and how can you tell? Use the terms and criteria that were presented in class.

   b. Discuss briefly the processes of erosion and deposition that are occurring in this area and why they are happening. What is the eventual fate of Oxbow Bend?

   c. Examine the river plain to determine if any river terraces are present. Describe the terraces you observe (i.e., size, how they formed), or suggest an explanation for why there are no terraces present.

2. Horseshoe Bend of the Colorado River, near Page (Arizona)
   Latitude: 36°52 40.99 N; Longitude: 111°30 36.43 W

   a. Explore this part of the Colorado River in Google Earth and describe the type of river present. How is this river similar or different to the portion of the Snake River you observed before?

   b. Examine the location of Horseshoe Bend. What type of feature is this? How is it similar and different to Oxbow Bend?

   c. Describe the Colorado River by zooming out: describe any terraces you observe (i.e., size, how they formed), or suggest an explanation for why there are no terraces present. What does this suggest about the geologic history of this part of Arizona compared to the portion of Wyoming observed earlier?

STREAM MORPHOLOGY CASE STUDIES, ON MARS

Using Google Earth, you can explore planet of the solar system. The location toolbar contains an icon that looks like Saturn: click the button to bring up a menu of locations—choose Mars.

Once Mars appears, you'll have a different set of layers to explore. The "Visible Imagery" contains the highest quality images, but the Viking Color Imagery layer is more uniform and may be easier to use in some places. As you explore the locations of this exercise, you might want to switch between datasets. To see the locations below, simply enter the name in the search string.

3. **Apollinaris Mons (also called Apollinaris Patera)**

   Zoom out so you can see the flanks of the volcano as well.

   a. Describe the linear features that surround Apollinaris Mons: if these were stream channels, what type of drainage would this be? Include a simple sketch of the drainage below.

4. **Orson Welles Crater**

Examine the valley that starts at the crater's NE rim known as **ShalbatanaVallis**.

a. Briefly describe this valley. Identify any evidence of erosion/deposition in the valley floor.

b. How does it differ from other channels you observed so far? Do you see any tributaries? This type of channel is called outflow channel and it is likely the result of the catastrophic floods that are thought to have formed this type of feature, on Earth, a similar feature is the Okavango Delta in Botswana, have you heard about this place?

Laboratory Experience Assessment for Stream Modification of Landscape

Learning Objective	Level of Confidence				
	Hesitant, concept is unclear, would not know how to use/apply	I have a general idea of what this is about and with guidance I could apply what I learned to problem solving		I am confident I understand this topic and I can apply it to solving a problem	
Describe the landscape in terms of uplands and valleys both in a map view and in cross section including measuring and estimating width and depth of the valleys in comparison to upland					
Recognize the evidence for youthful versus mature stream landscape from topographic maps and/or aerial views such as those in Google earth					
In general, how challenging is it for you to understand how landscape is modified by water by using maps?	Very challenging	Somewhat challenging, I can do it with some assistance		Not challenging	
What challenged/interested you the most about this activity? Why?					

Chapter 7

Flood Recurrence and Hazard

PRELAB STUDY SESSION

In preparation to this exercise, briefly answer the following questions. Refer to your textbook for the information you need. In addition to these questions, your lab instructor might give you more questions to answer.

A. Which factors lead to increased run-off and flooding?

B. What are the different types of flood? Why are there different types of flood?

C. Do floods recur regularly? What is the meaning and basis of flood recurrence intervals and flood prediction?

D. How are rivers monitored?

E. Why do people live in floodplains and what are the effects of floodplain development on flood gage height, flood lag time, flood recurrence?

STREAMFLOW CONDITIONS

1. The NOAA-USGS maintains a network of stations that monitor the streamflow conditions for streams across the Nation and alert about the flood conditions. Go to: http://water.weather.gov/ahps/
   a. Display data for locations near or above flood stage. Where are flood conditions occurring in the United States? Which major rivers are these conditions for?

   b. Click the warning and forecast tab. How does the flood gage correlates with the warning?

2. Hydrology conditions for the stream in the United States are monitored continuously and the data is available on the USGS Waterdata webpage at: https://waterdata.usgs.gov/nwis/rt Accotink Creek is a stream in Fairfax County, VA. Go to:
   https://waterdata.usgs.gov/usa/nwis/uv?01654000 or type in a search string: USGS 01654000 Accotink Creek. Go to the page and scroll to the graphs at answer the following questions:
   a. Summarize- the current discharge pattern and values as shown by the graph of discharge in cubic feet per second.

   b. How does that value compare to the median daily statistic for this stream? On how many years of observation is the mean calculated?

   c. How does the gage height correlate with the discharge? How does this correlation reflect weather conditions?

d. Local (County) Government informs citizen about flood hazard and best practices. A flood hazard map for Fairfax is available at
http://www.fairfaxcounty.gov/dpwes/stormwater/floodmaps.htm
a. Is our campus in a high flood hazard zone?

e. Local government issue best practices when flood conditions occur. For Fairfax county, go to: http://www.fairfaxcounty.gov/emergency/hazards/flooding.htm. What are you supposed to do in flood conditions?

f. Based on what you have learned about urbanization and flood and what you hear from local advisories, what is the most common type of flood in Fairfax County?

NOTE: Gaging stations are occasionally taken offline or decommissioned. These questions can be used for any other USGS monitored stream.

FLOOD RECURRENCE INTERVAL AND THE 100-YEAR FLOOD

1. Read http://water.usgs.gov/edu/100yearflood.html and answer the following questions:
   a. The origin of the term "one hundred year flood" is not a good representation of the actual meaning of this statistical measurement. What is a better way to describe it, from a hydrologist's viewpoint?

   b. Explain the advantage of having a long record of yearly river discharge measurements.

   c. Does a record breaking rainfall always cause a 100-year flood? Explain why or why not.

   d. At present there is no official recognition of flood recurrence intervals over 100 year. Here is a hydrologist viewpoint. Is a 500-year flood a conceivable recurrence interval?
   https://www2.usgs.gov/newsroom/docs/flooding_in_20080620.pdf

THE SUSQUEHANNA RIVER FLOOD RECURRENCE INTERVAL CASE STUDY

Calculating the recurrence interval is a matter of determining the relationship between gage height and discharge. For this exercise this will be done using the data from the spreadsheet.

Calculating the flood recurrence interval Plotting flood data:

> Using excel or equivalent software (or graph paper provided at the end of this chapter) generate an X-Y plot (points only) of gage height (feet) on the y-axis versus discharge (in cubic feet/second) on the x-axis.
>
> Label the plot axis (Y for gage height and X for discharge) and show the units of measurements (cubic feet/second). For ease of reading, keep a gridline on the plot. Use the best-fit line function through the data points. Save the plot as Chart 1.
>
> To determine the recurrence interval from the flood data available follow this procedure:
> - Sort the data by discharge, from highest to lowest values using the data/sort function. Create a rank column and assign the rank to the discharge starting with 1 for the highest discharge, 2 for the second highest, and so on.
> - Create a column to calculate the recurrence interval for each of the discharge value. Use the formula: RI = (n+1)/m where n = number of years of measurements (observations), and m = rank of the flood as assigned before (greatest = 1 to lowest = n).
> - Create another X-Y plot of discharge versus recurrence interval using the chart function. Plot the discharge axis (Y) using a *linear scale* and the recurrence interval on the X-axis using *a logarithmic scale*. Plot the best-fit curve (choose linear).

1. Answer the following questions.
   a. What is the recurrence interval for the 100 year flood?

   _____

   b. What is the 100-year flood discharge? Read the discharge for the 100-year flood on the second chart.
      100-year flood discharge _____

   _____

   c. What is the gage (stage) height for the 100-year flood? Read the value on the first graph you prepared _____

   _____

   d. Examine the dates of the high peak flood shown in Table 7.1. Which months have the highest recurrence of floods? Why?

   _____

Graph paper for discharge-gage height

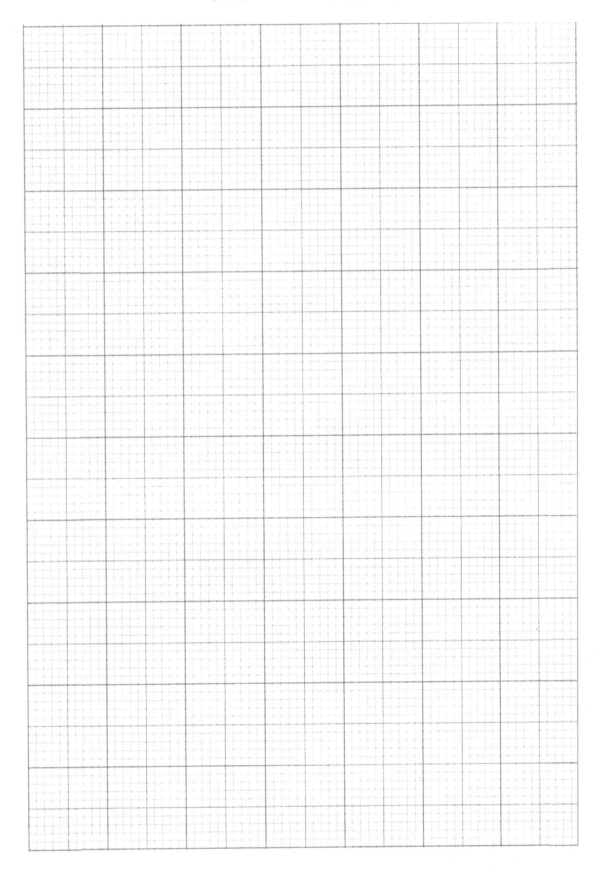

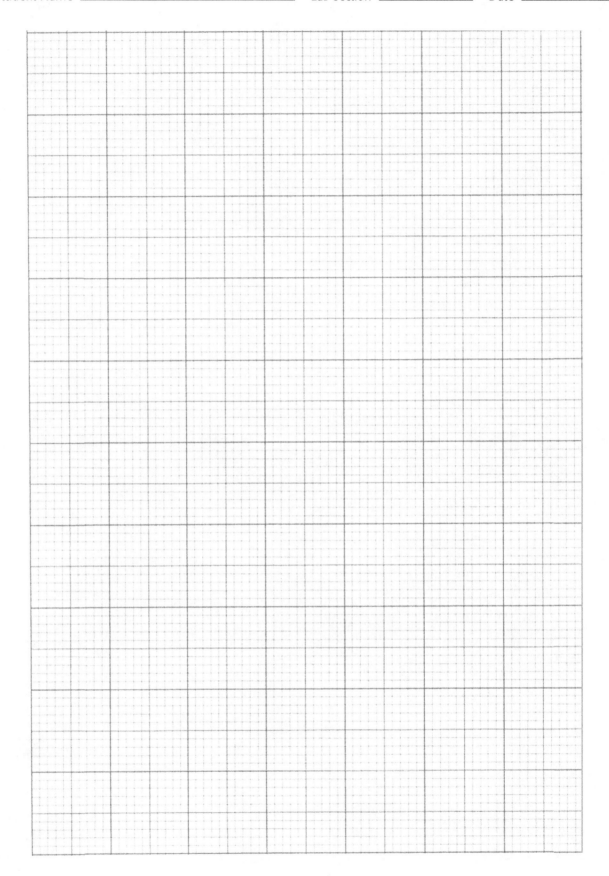

2. Mapping the 100-year flood hazard
   a. Using as a base map the Binghamton East topographic map (Figure 7.1), highlight the area that would be flooded by the 100-year flood. The elevation of the river gage at the town of Conklin is 840 feet. Add to this value the gage height for the 100-year flood, the sum of the two represents the height of the 100-year flood water. Highlight on the map the area that will be submerged by the flood. Use a pencil or a contrasting color and shade in the area.
   b. Describe in one paragraph the effect that a 50-year flood would have on the area and its population.

3. Current streamflow conditions the Susquehanna River at http://waterdata.usgs.gov/nwis/uv?01503000
   a. Scroll down to discharge and gage height plots shown in this page. What are the most recent conditions? Describe the trends shown in the graphs.

   b. How do this current trend compare with the average from the historical records? (compare to the triangular symbols)

   c. Summarize in a short paragraph how the stream flow over the last few weeks compares to the flood values used in the lab.

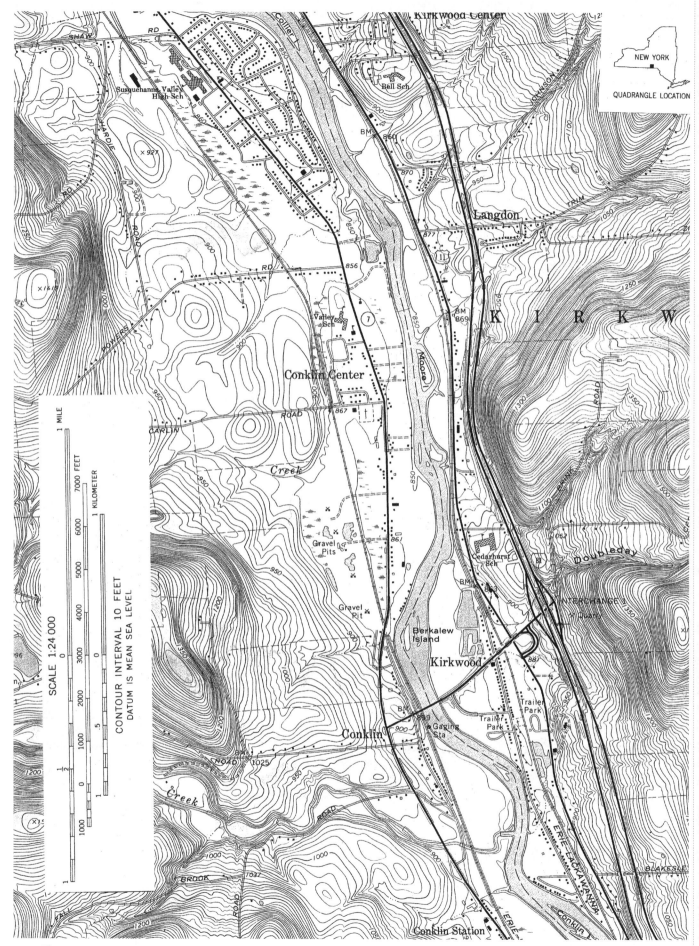

Figure 7.1   Binghampton East, NY Quad

Table 7.1

GAGE HEIGHT VS. DISCHARGE–HISTORICAL DATA

Year	Date	Gage Height (feet)	Discharge (cfs)	Year	Date	Gage Height (feet)	Discharge (cfs)
1913	3. 28, 1913	18.30	52,000[2]	1942	3. 19, 1942	14.45	28,100
1914	3. 30, 1914	18.00	47,000	1943	12. 31, 1942	18.76	48,600
1915	7. 08, 1915	16.15	40,500	1944	3. 18, 1944	14.80	30,000
1916	4. 02, 1916	16.49	42,100	1945	3. 18, 1945	14.17	27,500
1917	3. 28, 1917	13.54	28,700	1946	3. 09, 1946	15.49	32,900
1918	10. 30, 1917	13.73	29,400	1947	4. 06, 1947	15.04	31,000
1919	10. 31, 1918	10.65	17,900	1948	3. 22, 1948	20.83	60,500
1920	3. 29, 1920	15.05	35,200	1949	12. 31, 1948	14.39	28,400
1921	3. 10, 1921	13.17	27,100	1950	3. 29, 1950	15.87	34,600
1922	11. 29, 1921	16.03	39,900	1951	12. 04, 1950	16.20	36,100
1923	3. 24, 1923	13.23	27,300	1952	3. 12, 1952	13.40	24,700
1924	9. 30, 1924	16.86	44,000	1953	1. 25, 1953	13.61	25,400
1925	2. 12, 1925	17.04	44,900	1954	2. 18, 1954	14.55	29,000
1926	4. 10, 1926	14.04	30,600	1955	3. 13, 1955	12.72	22,500
1927	3. 15, 1927	14.81	33,600	1956	4. 07, 1956	16.04	39,200
1928	10. 19, 1927	16.88	43,500	1957	1. 23, 1957	11.74	21,400
1929	3. 17, 1929	17.60	47,000	1958	4. 07, 1958	15.83	38,300
1930	12. 20, 1929	10.90	18,600	1959	1. 22, 1959	14.49	32,300
1931	3. 30, 1931	12.16	22,800	1960	4. 06, 1960	17.02	44,000
1932	4. 01, 1932	13.75	29,000	1961	2. 26, 1961	16.02	39,100
1933	10. 08, 1932	13.10	25,000	1962	4. 01, 1962	15.17	35,300
1934	3. 05, 1934	13.20	25,400	1963	3. 28, 1963	15.73	37,800
1935	7. 09, 1935	16.95	41,900	1964	3. 10, 1964	18.26	50,200
1936	3. 18, 1936	20.14	61,600	1965	2. 10, 1965	9.81	14,900
1937	1. 26, 1937	12.88	24,300	1966	3. 06, 1966	10.68	18,000
1938	9. 23, 1938	15.89	34,100	1967	3. 30, 1967	10.30	16,800
1939	2. 21, 1939	15.64	33,100	1968	3. 23, 1968	11.63	21,200
1940	4. 01, 1940	19.13	51,800	1969	11. 19, 1968	10.53	24,000
1941	4. 06, 1941	13.40	24,900	1970	4. 03, 1970	12.54	25,300

Year	Date	Gage Height (feet)	Discharge (cfs)
1971	3. 16, 1971	11.51	21,700
1972	6. 23, 1972	12.89	26,500
1973	11. 09, 1972	14.40	32,100
1974	12. 28, 1973	12.43	24,900
1975	2. 25, 1975	14.05	30,700
1976	10. 19, 1975	14.31	31,700
1977	3. 16, 1977	16.90	43,400
1978	10. 19, 1977	16.28	40,300
1979	3. 07, 1979	17.25	45,200
1980	3. 22, 1980	12.59	25,400
1981	2. 21, 1981	12.39	24,700
1982	3. 27, 1982	10.31	17,700
1983	4. 16, 1983	13.84	29,800
1984	12. 14, 1983	17.17	44,700
1985	9. 28, 1985	11.04	20,000
1986	3. 15, 1986	17.10	44,400
1987	11. 27, 1986	12.50	25,100
1988	5. 20, 1988	11.49	21,500
1989	5. 07, 1989	12.48	25,000
1990	2. 17, 1990	11.12	20,300
1991	10. 24, 1990	12.18	24,000
1992	3. 12, 1992	9.46	15,100
1993	4. 01, 1993	17.91	48,500

Year	Date	Gage Height (feet)	Discharge (cfs)
1994	4. 07, 1994	13.42	28,300
1995	3. 09, 1995	9.63	15,600
1996	1. 19, 1996	17.55	46,600[2,9]
1997	12. 02, 1996	14.29	31,600
1998	1. 10, 1998	15.42	36,400
1999	1. 24, 1999	14.89	34,100
2000	2. 28, 2000	15.78	38,000
2001	4. 11, 2001	13.58	28,900
2002	3. 27, 2002	12.09	23,700
2003	3. 23, 2003	14.73	33,500
2004	9. 18, 2004	19.01	54,700
2005	4. 03, 2005	18.08	49,400
2006	6. 28, 2006	25.02	76,800
2007	3. 28, 2007	12.64	25,100
2008	3. 09, 2008	14.26	30,700
2009	3. 11, 2009	12.33	24,100
2010	1. 25, 2010	13.39	27,600
2011	9. 08, 2011	23.94	72,100
2012	1. 28, 2012	9.08	15,000
2013	6. 29, 2013	11.15	20,300
2014	5. 17, 2014	13.4	23,300
2015	4. 10, 2015	12.97	26,300
2016	Feb. 26, 2016	10.87	19,400

Note:
- [2]—Discharge is an estimate
- [9]—Discharge due to snowmelt, hurricane, ice-jam or debris dam breakup

FLOOD HAZARDS IN THE WORLD

1. Go the Dartmouth Flood Observatory Website. The Flood observatory collects and archives in near real time global remote sensing measurements and mapping of flood information. Go to: http://floodobservatory.colorado.edu/

   a. Scroll below the featured flood and examine the world map showing the ongoing flooding in the world (each event has an identifier number). Which areas of the world are flooding? (Note: there is an approximate two week lag before most of the values are updated)

   b. On the left menu of the map, click Current events. A table with the recently archived events will appear. Notice that the columns report several descriptors of each event (register #.): Country, location details, date of beginning and end, area affected, main cause, number of people displaced, and so on.
   Using the data available for the 12 most recent floods, prepare a presentation quality table that ranks these recent floods based on criteria of your choice that you consider important in flood hazard assessment. The table must have a maximum of 10 columns, so choose what you want to display. You can do this with a spreadsheet or you can use the blank table next page.

   c. Compare and contrast the flooding events. What most common process triggered these floods?

   d. Plot on the world map the floods discussed in this exercise 6.

Student Name _____ Lab section _____ Date _____

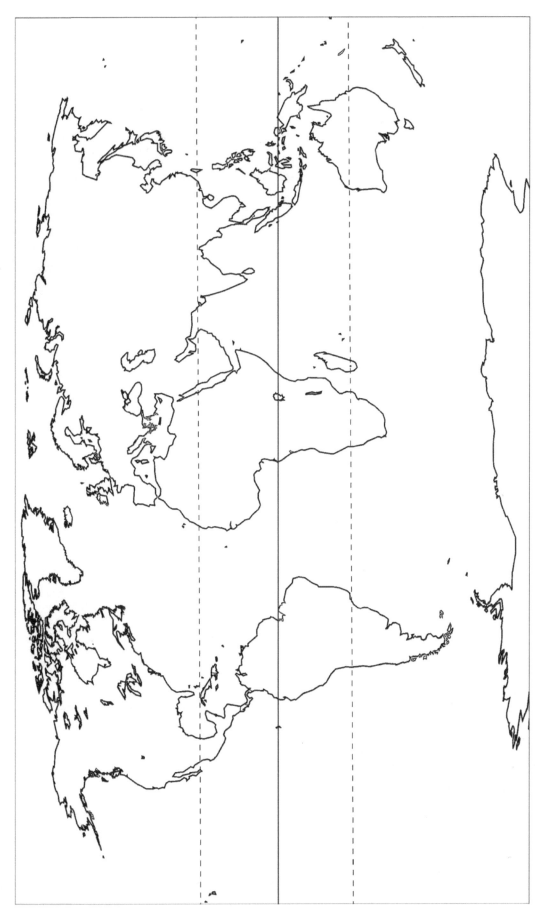

Laboratory Experience Assessment: Flooding

Learning Objective	Level of Confidence		
	Hesitant, concept is unclear, would not know how to use/apply	I have a general idea of what this is about and with guidance I could apply what I learned to problem solving	I am confident I understand this topic and I can apply it to solving a problem
Understand the meaning of stream data (discharge, gauge heights, recurrence interval, etc.) and their range of variability.			
Work with the data on the spreadsheet, organizing and plotting data and formatting charts to achieve a meaningful representation of data.			
Understand the meaning and limitations of the Recurrence Interval and the 100 year flood.			
Finding data related to this and similar problems on line from the USGS website.			
How challenging was this exercise for you?			
What challenged/interested you the most about this activity? Why?			

Chapter 8

Coastline Evolution

PRELAB STUDY SESSION

In preparation to this exercise, briefly answer the following questions. Refer to your textbook for the information you need. In addition to these questions, your lab instructor might give you more questions to answer.

A. Explain how wave refraction occurs (i.e., how the waves break roughly parallel to the shoreline).

B. Which coast in the United States is predominantly composed of submergent coastlines? How can you tell? Give specifics topographic clues as to how you came up with your answer.

C. Sketch the dominant direction of the longshore current along the eastern seaboard.

D. Now that you are thinking of the east coast shoreline, is it primarily an area of erosion or deposition? Be specific as to how you came up with your answer.

E. What are jetties, groins, and seawalls? Why are they beneficial?

HUMAN MODIFICATION OF SHORELINES CASE STUDY 1: OCEAN CITY, MARYLAND

Ocean City is located on a long, narrow barrier island called Fenwick Island. This area was the site of small fishing villages since the 18th century. During the 19th century it became a vacation resort open only in the summer. In August 1933, a severe Nor'easter* ripped open the barrier island and tidal currents swept across the gap, separating the barrier island into a northern section (Fenwick Island) and a southern section (Assateague Island).

The newly opened seaway to the Atlantic was soon stabilized with jetties and seawalls. It was then that Fenwick Island was developed and Ocean City became a popular year-round vacation destination. The southern island, Assateague, remained undeveloped and it is now a National Seashore.

Carefully observe the maps and the aerial photos shown in Figures 8.1 to 8.6. After the 1933 storm, the Army Corps of Engineers constructed jetties on either side of the inlet to keep it open. The southern jetty is labeled Seawall on the map. Answer the following questions:

1. Based on observation of the maps and images, describe the orientation and the direction of movement of longshore currents. Justify your answer.

2. How does the longshore current relate to the Gulf Stream and the general Atlantic Circulation? See http://oceancurrents.rsmas.miami.edu/index.html. Click on Atlantic Ocean and then Gulf of Mexico.

3. Notice that Assateague Island has migrated west from its position in 1901. This migration began in 1933. Why did this occur?

*A Nor'easter is a cyclonic storm that moves along the east coast of North America. They may occur any time of year, but are most frequent in the late summer.

4. Groins have been constructed on the eastern side of Fenwick Island. What effect do these groins have on the beaches of Ocean City Municipal Pier and why?

5. The vertical aerial photographs of the Ocean City Inlet were taken in 1938 and 1952 (Figures 8.2 and 8.4). What is the actual area covered by these photographs? Calculate the area and show work in the International System of Units (SI).

6. Note the changes that are evident through the images as you examine them in a temporal sequence. Open Google Earth and "fly" to the area covered by the maps. Summarize the natural and manmade changes to the barrier island, the inlet, and Ocean City that you observe.

7. The westward migration of Assateague Island could be reversed if the groins and jetties of Fenwick Island were removed. Would the removal of these structures put properties in Ocean City at greater risk? Justify your answer.

8. Read the short article by Williams (USGS, 2002) on the effects of Ocean City Inlet on Assateague Island: http://soundwaves.usgs.gov/2002/11/research.html

   How far south have the effects of Ocean City inlet been observed?

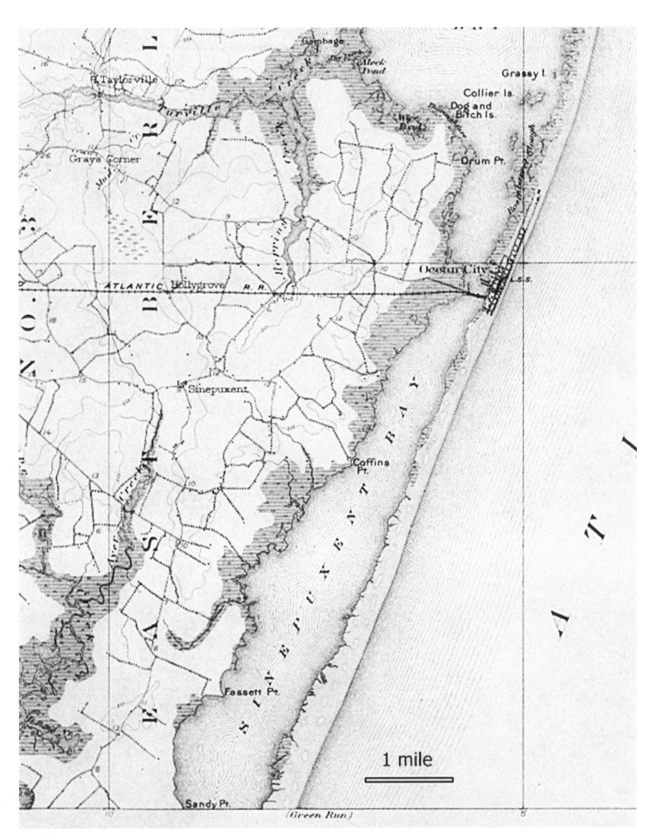

Figure 8.1 Ocean City, Maryland (1901).
detail of USGS Map

Figure 8.2 Vertical aerial photograph of Ocean City Inlet, Maryland taken in 1938. Photograph scale is approximately 1:20,000. North is toward the top of the page.

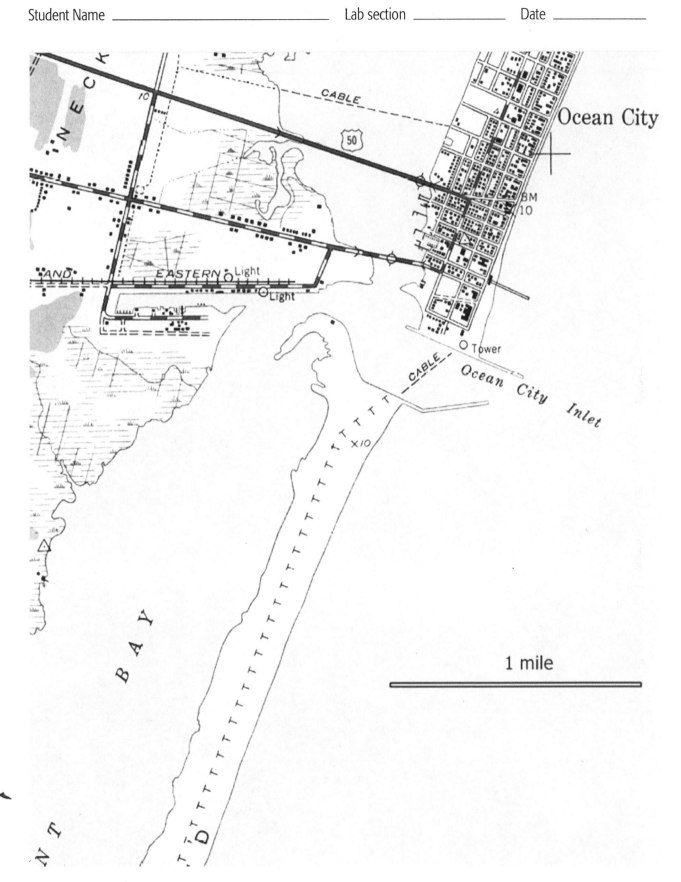

Figure 8.3   Ocean City, Maryland (1942).
detail from 7.5 × 7.5 USGS map

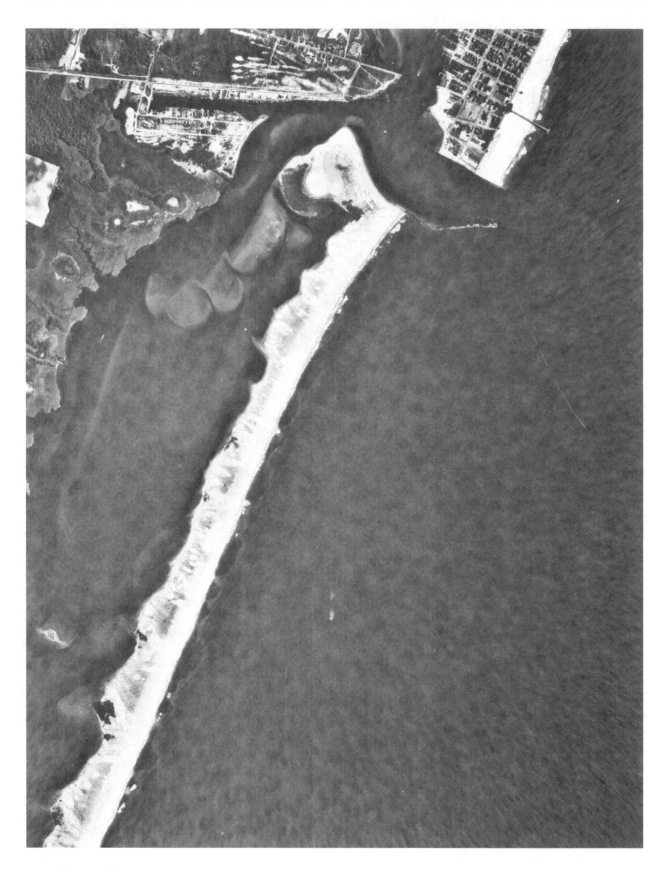

Figure 8.4   Vertical aerial photograph of Ocean City Inlet, Maryland taken in 1952. Photograph scale is approximately 1:20,000. North is toward the top of the page.

Student Name ______________________ Lab section ____________ Date ____________

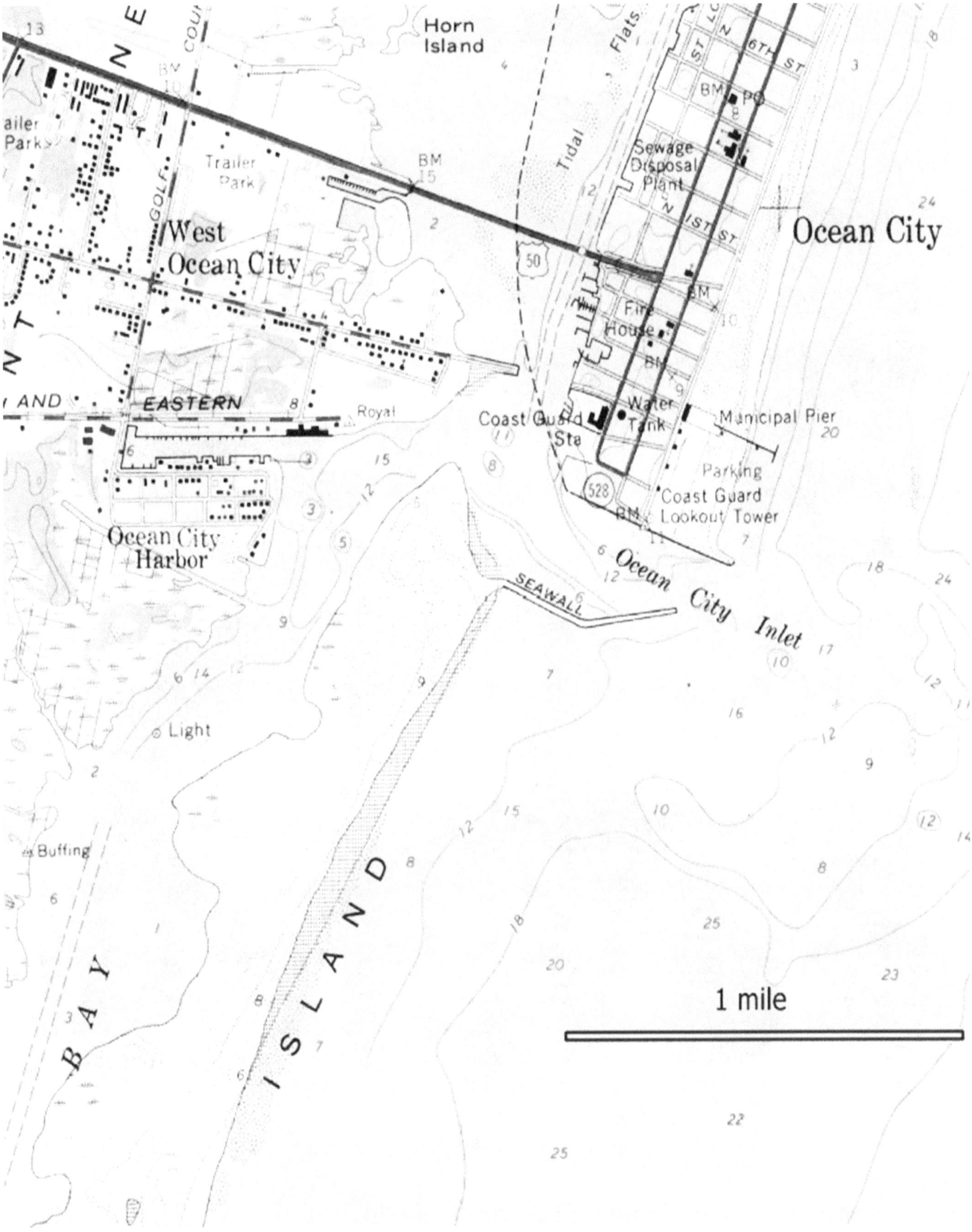

Figure 8.5 Ocean City, Maryland (1964).
detail of 7.5 × 7.5 USGS map

Figure 8.6   Ocean City, Maryland (2014).
Satelite Image

HUMAN MODIFICATION OF SHORELINES CASE STUDY 2: NEW JERSEY COAST

1. In October 2012, Hurricane Sandy slammed into the New Jersey coast. This storm was the second most destructive storm in United States history and affected 24 states including the entire eastern seaboard from Florida to Maine, the Appalachian Mountains, and as far inland as Michigan and Wisconsin. Go to the NOAA National Geodetic Survey web page for Hurricane Sandy: http://oceanservice.noaa.gov/news/ weeklynews/nov12/ngs-sandy-imagery.html.

   Examine the high-resolution photos on the page. Note that by scrolling across the images you will be able to see the same area before and after the passage of Sandy.

   a. Compare the before and after pictures of the New Jersey shore and describe the changes in sediment erosion, transportation, and deposition caused by the storm surge. Make a general statement after observing all the images on the webpage.

   b. How did the structures engineered to protect the coastline respond to the storm?

NATURAL EVOLUTION OF SHORELINE CASE STUDY 3: CHESAPEAKE BAY

Refer to what you learned in lecture about coastal processes to answer the following questions.

1. Look at the Washington, D.C., Maryland, Virginia topographic map (Figure 8.8). This area is typical of the Chesapeake Bay region and much of the central Atlantic coast of the United States. All major areas of water are interconnected with the sea and are affected by tidal fluctuations.

    a. Is this an emergent or a submergent coast? Discuss the topographic features which enable you to make this distinction.

2. What appears to have been the active erosional agent prior to the most recent change in sea level? On what criteria do you base your answer?

3. A topographic profile is provided in Figure 8.9.
    Indicate on your profile those land areas which have been affected from earlier erosional agents and those which have been produced by shoreline processes.

4. What is the minimum estimate of the vertical extent to which sea level has recently been changed along this coast? Explain your calculations.

5. What geomorphic name is applied to the Potomac River in this section?

Adpated pages 293–294 from *Physical Geology: Laboratory Text and Manual, 6th Edition* by R.D. Dallmeyer. Copyright © 2000 by Kendall Hunt Publishing Co. Reprinted by permission.

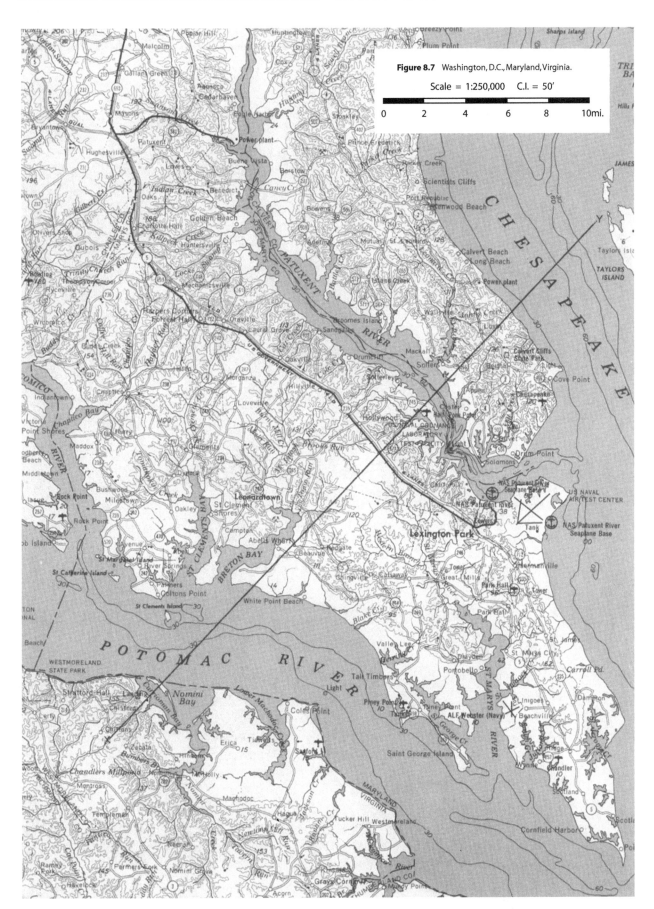

Figure 8.7 Washington, D.C., Maryland, Virginia.

Scale = 1:250,000   C.I. = 50′

0   2   4   6   8   10mi.

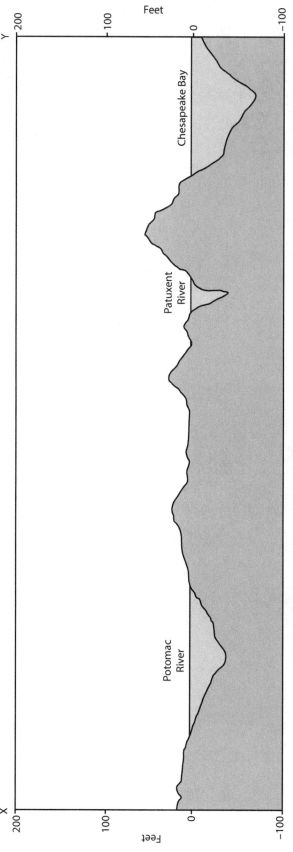

Figure 8.8 Topographic profile: Washington, D.C., Maryland, Virginia (VE = 208×).

Laboratory Experience Assessment: Coastlines Evolution

Learning Objective	Level of Confidence		
	Hesitant, concept is unclear, would not know how to use/apply	I have a general idea of what this is about and with guidance I could apply what I learned to problem solving	I am confident I understand this topic and I can apply it to solving a problem
Use evidence from topographic maps and imagery to identify the action of longshore currents and drift			
Recognize submergent and emergent coastlines and their distinctive features			
Identify and explain the effect of engineered structures along coastlines			
Appreciate the pros and cons of beach replenishment			
How challenging did you find this exercise?			
What challenged/interested you the most about this activity? Why?			

Tectonics

Chapter 9

Volcanic Processes

PRELAB STUDY SESSION

In preparation to this exercise, briefly answer the following questions. Refer to your textbook for the information you need. In addition to these questions, your lab instructor might give you more questions to answer.

    A. List and sketch the various type of volcanoes. Sketch them to scale with respect to each other.

    B. What materials can a volcano erupt? It is not just lava. List volcanic products by composition and by type (tephra, size, volcanic gases, etc.).

    C. Name some possible volcanic hazards.

    D. The differences in volcanic behavior are due to the combination of many elements such as magma chemistry, volatiles in the magma, type of volcanic structure, and tectonic settings. How can we study what volcanoes are doing? How can we deduce out what a volcano will do?

MAGNITUDE OF VOLCANIC ERUPTIONS

Volcanic Explosivity Index (VEI) provides a relative measure of the explosiveness of a volcanic eruption; it was defined by the volcanologists Newhall and Self in 1982. This index is calculated by including the volume of products, the eruption cloud height, and qualitative observations on the magnitude of the eruption The scale is open-ended, but the largest volcanic eruptions known have been assigned a VEI index of 8 (Yellowstone eruption, ejected 1,000,000,000,000 m^3 of tephra and had a cloud column height of over 50 km). A value of 0 is given for nonexplosive eruptions, defined as less than 10,000m^3 of tephra (Table 9.1).

The names under the classification originate from well-known volcanoes that best represent varying types of eruption, Hawaii for voluminous, mostly effusive eruptions, Stromboli and Vulcano (two volcanoes in the Mediterranean region). Plinian and ultraPlinian come from the Roman author Pliny the younger, who described in great detail the eruption of Mount Vesuvius in 79 AD.

Table 9.1   Volcano Explosivity Index

V.E.I.	Description	Plume Height (km)	Volume (m³)	Classification	Example
0	Non-explosive	< 0.1	1000s	Hawaiian	Kilauea
1	Gentle	0.1–1	10,000s	Hawaiian/Strombolian	Stromboli
2	Explosive	1–5	1×10^6	Strombolian/Vulcanian	Galeras, 1992
3	Severe	3–15	10×10^6	Vulcanian	Ruiz, 1985
4	Cataclysmic	10–25	100×10^6	Vulcanian/Plinian	Galunggung, 1982
5	Paroxymal	> 25	1 km^3	Plinian	St. Helens, 1980
6	Colossal		10s km^3	Plinian/Ultraplinian	Krakatau, 1883
7	Super-colossal		100s km^3	Ultraplinian	Tambora, 1815
8	Mega-colossal		1,000s km^3		Yellowstone, 2Ma

1. Do you know where the example volcanoes in Table 9.1 are located on Earth? Map their position on the world chart at the end of this chapter. Add the location of Vesuvius and the Eolian Island volcanoes.

CASE STUDY: MOUNT ST. HELENS

Mount St. Helens (MSH) erupted in 1980. MSH is a composite volcano built from layer of lavas and pyroclastic flows ejected during the volcanos many explosive cycles that characterize its complex geologic history.

The volcano formed during nine eruptive periods beginning about 275,000 years ago. Mapping of rocks and structures in the area indicated that tephra, lava domes, and pyroclastic flows formed an older St. Helens edifice. The modern edifice has been building up during the last 3000 years by flows erupting from the summit and flank vents, ranging in composition from basalt to andesite. Eruptions observed in the nineteenth century originated from the north flank.

Prior to the 1980 eruption, MSH had a conical, well-defined shape. During the 1980 eruption the upper 400 m of the summit was removed by slope failure, leaving a 2 × 3.5 km horseshoe-shaped crater now partially filled by a lava dome.

Go to the Cascade Volcano Observatory website to see images of MSH before and after the eruption: http://volcanoes.usgs.gov/volcanoes/st_helens/st_helens_multimedia_gallery.html

Table 9.2 summarizes the eruptive history of MSH starting from the eruptions that crop out at the base of the "recent" MSH. The eruptive episodes are grouped by duration and eruptive characteristics. The table is based on data compiled by the Global Volcanism Program of the Smithsonian Institution from published scientific literature on MSH. Starting dates of eruptions were calculated by dating techniques such as radiocarbon, tephrochronology, and dendrochronology. Historical records are available only for the most recent eruptive episodes. When possible the VEI was calculated.

2. Using the data available in Table 9.2 to answer the following questions:
   a. What are the common traits of MSH eruptions with respect to the VEI, type of eruptions, site of eruption, volcanic materials erupted, and mode of emplacement (e.g., type of flows)? Which ones are most common? Summarize your observations.

   b. Does MSH eruptive history suggest that the 1980 eruption was to be expected? What other visual and geophysical monitoring helped forecast the eruption? Read the paragraph "precursory activity" at: http://volcanoes.usgs.gov/volcanoes/st_helens/st_helens_geo_hist_99.html
   What were the most compelling evidence that the eruption was imminent?

   c. Which one was the largest eruption on record?

Table 9.2   Eruptive History of Mount St. Helens, Compiled and Simplified from St. Helens Eruptive History—Global Volcanism Program
http://volcano.si.edu/world/volcano.cfm?vnum=1201-05-&volpage=erupt

October 2004–January 2008

VEI: 2–Lava Volume: $> 9.3 \times 10^7 \, m^3$
- Explosive eruption and phreatic explosion(s) from central vent and flank vent
- Lava dome extrusion and spine extrusion
  Mudflow(s) (lahars)

November 1990–February 1991

VEI: 3
- Explosive eruption and Phreatic explosion(s) from central vent
- Pyroclastic flow(s) + Mudflow(s) (lahars)

December 1989–January 1990

VEI: 2
- Explosive eruption and Phreatic explosion(s) from central vent

March 1980–October 1986

VEI: 5–Lava Volume: $7.4 \times 10^7 \, m^3$–Tephra Volume: $1.2 \times 10^9 \, m^3$
- Explosive eruption and Phreatic explosion(s) from central vent and flank vent
- Pyroclastic flow(s); Lava dome extrusion + spine extrusion
- Mudflow(s) (lahars) + Debris avalanche(s)

1847–1857

VEI: 2
- Intermittent episode of explosive eruption from flank vent
- Lava dome extrusion (1847 only)

1842–1845

VEI: 3–Tephra Volume: $1 \times 10^7 \, m^3$
- Explosive eruption from flank vent
- Pyroclastic flow(s); Lava dome extrusion

March 1835

VEI: 2
- Explosive eruption from flank vent
- Lava flow(s); Lava dome extrusion

August 1831

VEI: 3
- Explosive eruption from flank vent
- Lava flow(s); Lava dome extrusion (?)

January–March 1800

VEI: 5–Tephra Volume: $1.5 \times 10^9 \, m^3$
- Explosive eruption from flank vent
- Lava flow(s); Mudflow(s) (lahars)

1610 ± 40 years

VEI not available
- Explosive eruption from central vent
- Pyroclastic flow(s); Mudflow(s) (lahars); Lava dome extrusion

1525 ± 25 years

VEI not available
- Explosive eruption from central vent
- Pyroclastic flow(s), Mudflow(s) (lahars); Lava dome extrusion

1482

VEI: 5–Tephra Volume: $1.5 \times 10^9 \, m^3$
- Explosive eruption from central vent

Table 9.2 *Continued.*

1480

VEI: 5+–Tephra Volume: $7.7 \times 10^9 \text{ m}^3$
- Explosive eruption from central vent
- Pyroclastic flow(s), Mudflow(s) (lahars); Lava dome extrusion

780 AD ± 300 years

VEI not available
- Explosive eruption from flank vent
- Pyroclastic flow(s), Mudflow(s) (lahars); Lava dome extrusion

420 AD; 270 AD isolated eruptive phases

VEI not available
- Explosive eruption from central vent
- Lava flow(s), Mudflow(s) (lahars)

230 AD

VEI: 0
- Lava flow(s)

190 AD (?)

VEI not available
- Explosive eruption from central vent
- Lava dome extrusion

100 AD (?)

VEI: 0
- Explosive eruption from flank vent
- Lava flow(s)

100 BC; 220 BC; 250 BC; 280 BC isolated eruptive phases

VEI not available
- Explosive eruption from central vent
- Pyroclastic flow(s); Lava flow(s); Mudflow(s) (lahars); Lava dome extrusion

530 BC (?)

VEI: 5–Tephra Volume: $1.2 \times 10^9 \text{ m}^3$
- Explosive eruption
- Pyroclastic flow(s), Mudflow(s) (lahars); Lava dome extrusion

830 BC ± 75 years; 1180–1010 BC; 1610–1680 BC

VEI not available
- Explosive eruption
- Pyroclastic flow(s); Mudflow(s) (lahars); Lava dome extrusion

1770 BC ± 100 years

VEI: 5–Tephra Volume: $3.5 \times 10^9 \text{ m}^3$
- Explosive eruption
- Pyroclastic flow(s); Lava dome extrusion

1860 BC (?)

VEI: 6–Tephra Volume: $1.5 \times 10^{10} \text{ m}^3$
- Explosive eruption
- Pyroclastic flow(s); Mudflow(s) (lahars); Lava dome extrusion

2100 BC ± 300 years

VEI not available
- Explosive eruption

2340 BC (?)

VEI:5–Tephra Volume: $1.2 \times 10^9 \text{ m}^3$
- Explosive eruption
- Pyroclastic flow(s)

3. The eruption of May 18, 1980 produced dramatic changes in the shape and volume of MSH using the preeruption topographic maps (Figure 9.2) build a North-South topographic profile that cuts across the summit (9677 ft). Use only the index contour lines (CI 400 ft). Construct a second topographic profile along the same line using the posteruption map (Figure 9.3). Note that this map has index contour at 200 CI. You can build the profiles on the same graph so they will overlap.

   a. Compare the two profiles describe what changes have occurred along the profile

4. MSH before the eruption had a regular cone shape, with a base-diameter of about 6 km and an elevation of about 3 km from the surrounding area. The preeruption volume of the volcano can be approximated to that of a cone:

   $$V = 1/3 \; \theta \; R^2 h \text{ (where R is the radius and h is the height).}$$

   a. Using the formula for the volume of the cone, calculate the volume of pre-1980 MSH in cubic kilometers. Show your calculations.

   b. The 1980 eruption removed about 6.5 km3 of the volcanic edifice, what percentage of the volcano does was then removed by the eruption? Show your calculations.

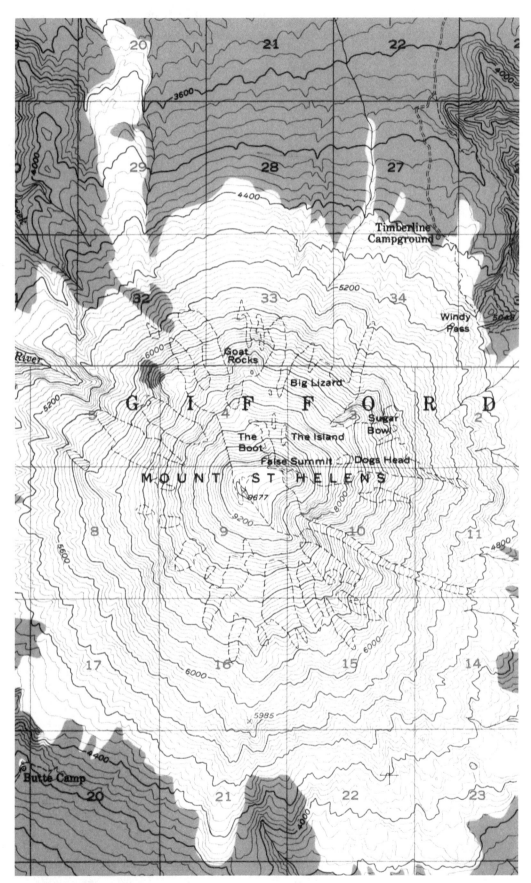

Figure 9.1   Topographic map of Mount St. Helens before the eruption. Scale 1:50,000.

Figure 9.2 Topographic map of Mount St. Helens after the eruption. Scale 1:50,000.

c. Convert the total volume of lava + tephra erupted in 1980 in cubic kilometers (see Table 9.2). What is the total volume of material produced by the eruption (Note: This is Juvenile material, freshly produced by the volcano!)

d. How does the 1980 eruption of juvenile material compares to the other eruptions of MSH (see Table 9.2).

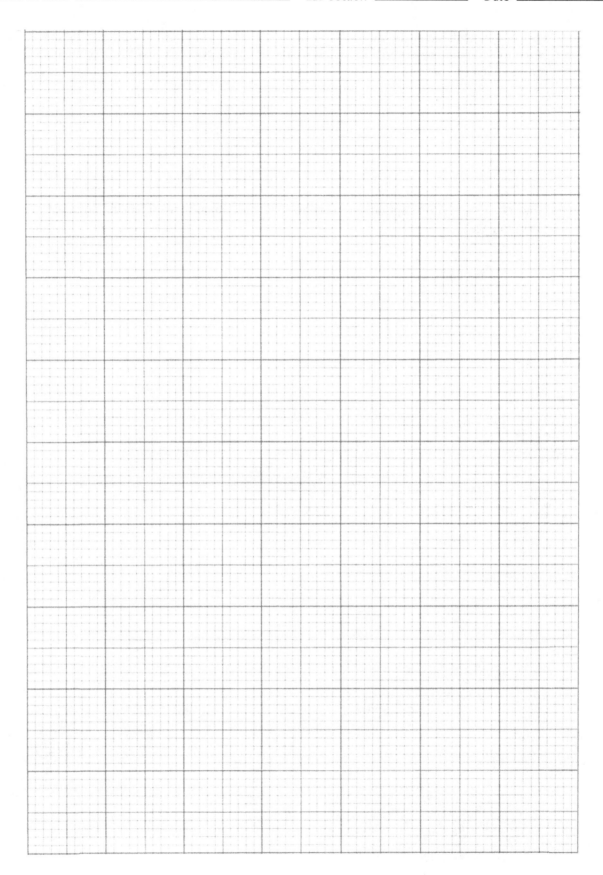

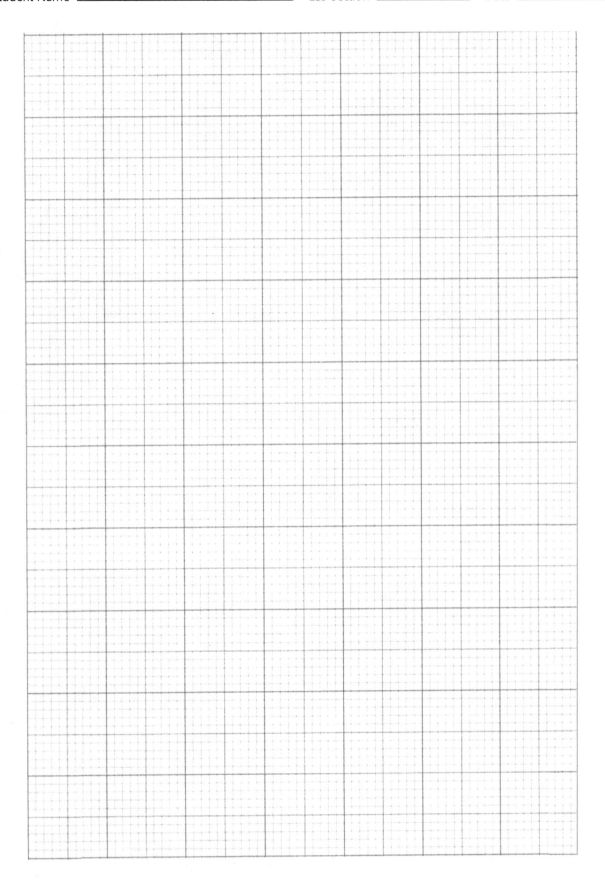

5. Like any other active volcano, MSH needs to be closely monitored in order to assess the level of activity and the likelihood of an eruption. To see what physical parameters are been monitored at MSH go to: http://volcanoes.usgs.gov/volcanoes/st_helens/st_helens_monitoring_16.html

   a. List the physical parameters that are monitored by Cascade Volcanological Observatory. Which ones need sensors on the ground, which ones are detected by remote sensing?

   b. What was the depth of the earthquakes in 1980? When were earthquakes at that depth observed again?

   c. What is the other typical depth of earthquakes at MSH observed during the 1980–2012 interval and what does it indicate?

   d. Return to the main monitoring page for MSH. Scroll below the instrument map to see the depth of the earthquakes recorded this week. What is the number of earthquakes observed, what is their depth?

   e. What deformation of the ground characterized the 2004–2008 eruptive episode of MSH and what did it indicate about the internal processes of the volcano?

   f. What does the volcanic gas SO2 indicate about the movement of magma?

   g. What trends in SO2 emission have been observed at MSH?

VOLCANIC HAZARDS

Volcanic activity may hinder or compromise human life or interest, so volcanic activity is often recognized as a geologic hazard. The United States Geological Survey (USGS) runs the volcanological observatories of the United States and it is part of the World Organization of Volcanological Observatories.

The international community recognizes found levels of alert that are meant to inform people on the ground about volcanic activity. The alert levels are issued in conjunction with the aviation alert code because aircraft is especially vulnerable when flying into dispersed volcanic ash, virtually invisible to the radar onboard. Alert codes are summarized in the table below:

	Normal	Advisory	Watch	Warning
Volcano Alert Level Land Based	Volcano is in typical background, non-eruptive state or, *after a change from a higher level,* volcanic activity has ceased and volcano has returned to non-eruptive background state.	Volcano is exhibiting signs of elevated unrest above known background level or, *after a change from a higher level,* volcanic activity has decreased significantly but continues to be closely monitored for possible renewed increase.	Volcano is exhibiting heightened or escalating unrest with increased potential of eruption, timeframe uncertain, **OR** eruption is underway but poses limited hazards.	Hazardous eruption is imminent, underway, or suspected.
Aviation Color Codes	If possible Specify Plume height			
	Green	Yellow	Orange	Red

6. Visit the USGS Volcano Hazard page at: http://volcanoes.usgs.gov/. Answer the following questions:

    a. How many volcanoes are reported on Red alert? _____ on Orange? _____ How many on Yellow?_____

    b. Where are these volcanoes located? Use the mute map of the World to mark the position of the volcanoes currently on high level of alert.

VOLCANOES OF THE WORLD

The Global Volcanism Program of the Smithsonian Institution in collaboration with the USGS publishes a weekly bulletin of ongoing volcanic activity. Visit the page:

http://volcano.si.edu/reports_weekly.cfm

7. Choose a volcano from those actively erupting this week. Make sure it is a volcano you never heard about before (learn something new!) and it was not mentioned previously in this exercise. Click on the symbol to access the volcano info page and answer the following questions.

  a. What is "your" volcano name?

  b. Where is it located? (region and Nation in the World). Plot the position of your volcano on the map at the end of the chapter.

  c. What type of volcano is this? (summarize the description)

  d. What type of eruptions are typical of your volcano?

  e. What is its highest VEI on record?

  f. When did it erupt last time?

g. Are other volcanoes found in the same region? Why is this volcano there? (a plate boundary, a hot spot? etc.)

h. Go to: http://si-vmarcgis01.si.edu/thisdynamicplanet/ Give time to the app to load and then look for the position of the volcanoes that have erupted since 1900. These are active volcanoes that might not be erupting now. Choose one in an area that you didn't expect to have volcanoes. Plot its position and name on the world map.

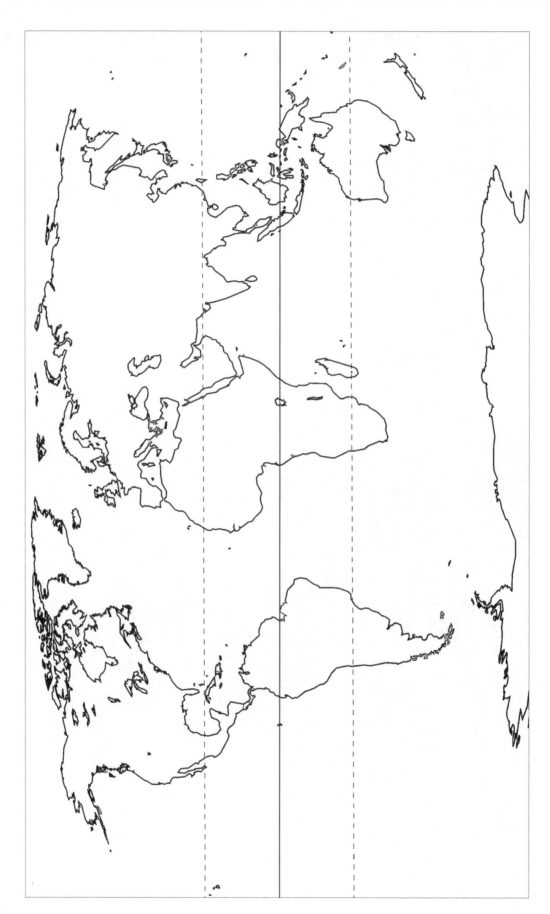

Laboratory Experience Assessment: Volcanic Processes—The Mt. St. Helens Case Study

Learning Objective	Level of Confidence		
	Hesitant, concept is unclear, would not know how to use/apply	I have a general idea of what this is about and with guidance I could apply what I learned to problem solving	I am confident I understand this topic and I can apply it to solving a problem
Evaluate data collected from diverse sources (historical, field evidence, diverse dating techniques, V.E.I.) in describing eruption history of Mount St. Helens (or any other volcano)			
From field measurements estimate quantitatively the changes in the landscape (topographic profiles before and after eruption, and calculation of volume)			
Appreciate the significance and supporting evidence related to assessment of natural hazards			
How challenging did you find this exercise?			
What challenged/interested you the most about this activity? Why?			

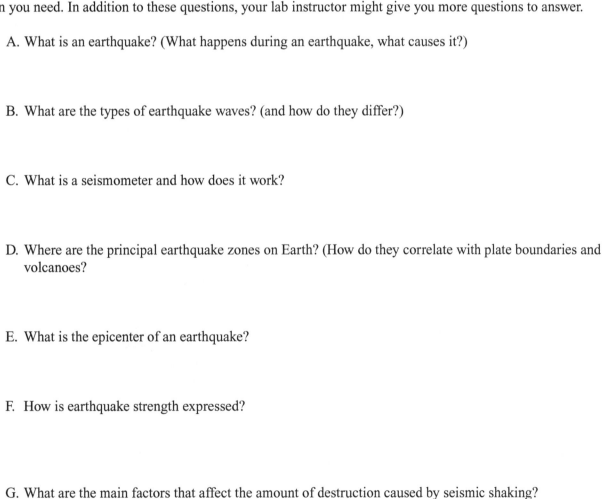

Chapter 10

Earthquake Analysis

PRELAB STUDY SESSION

In preparation to this exercise, briefly answer the following questions. Refer to your textbook for the information you need. In addition to these questions, your lab instructor might give you more questions to answer.

A. What is an earthquake? (What happens during an earthquake, what causes it?)

B. What are the types of earthquake waves? (and how do they differ?)

C. What is a seismometer and how does it work?

D. Where are the principal earthquake zones on Earth? (How do they correlate with plate boundaries and volcanoes?

E. What is the epicenter of an earthquake?

F. How is earthquake strength expressed?

G. What are the main factors that affect the amount of destruction caused by seismic shaking?

TRAVEL TIME CURVES

At any given seismic station, the lag time between the arrival of the P and S waves is used to measure the distance between the recording station and the epicenter. This relationship is expressed in the travel-time graph (Figure 10.1).

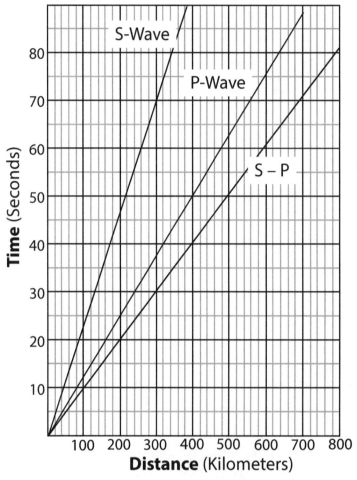

Figure 10.1   Travel-time graph recording the time for P and S waves

Examine the travel-time graph of Figure 10.1. The upper curve shows the travel-time for the S waves, the center one shows P waves travel-time. The lower curve shows the lag (time difference in arrival) between the S and P waves as a function of the distance between the epicenter and the seismic station. You can use the travel time also to figure out when the seismic waves started to travel away from the epicenter by subtracting the S-P lag time from the arrival time at any given station.

Check your understanding of travel-time graph:
1. What is the approximate distance between the epicenter and a seismic station that records a S-P travel time difference (lag) of 60 seconds?

2. The seismic station on the previous question recorder the P waves first arrival precisely at 8:15 a.m. When did the earthquake strike the epicenter?

Student Name _____ Lab section _____ Date _____

DETERMINING THE EPICENTER OF AN EARTHQUAKE

http://www.sciencecourseware.com/VirtualEarthquake/

Examine the three seismograms of an earthquake as they were recorded by different seismic stations (Figures 10.2 to 10.4). For each seismogram you will measure the S-P time interval (in seconds). The S-P time interval is directly related to the distance the waves have traveled from the earthquake's point of origin to a station. (Figure 10.1) The actual location of the earthquake's epicenter will be on the perimeter of a circle drawn around the recording station. The radius of this circle is called the epicentral distance; the direction which the waves came from is unknown. Three stations are needed in order to "triangulate" the location. Use the seismograms below to estimate the S-P time for each of the recording stations:

1. **Fresno, CA Seismic Station**

   S-P Interval (sec) = _____

   Epicentral distance (km) = _____

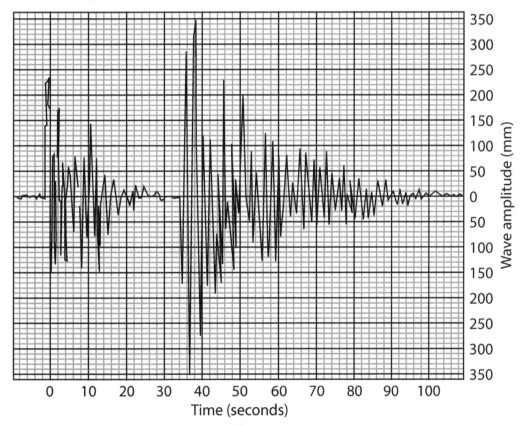

Figure 10.2

2. Las Vegas, NV Seismic Station

    a. S-P Interval (sec) = _____

    b. Epicentral distance (km) = _____

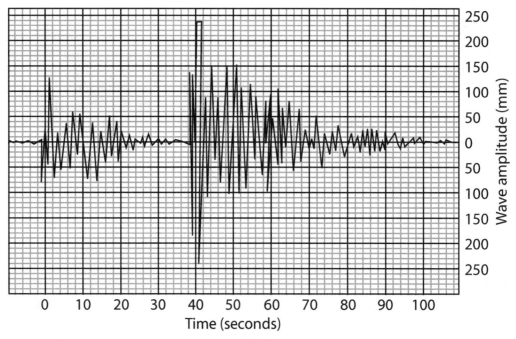

Figure 10.3

3. Phoenix, AZ Seismic Station

    a. S-P Interval (sec) = _____

    b. Epicentral distance (km) = _____

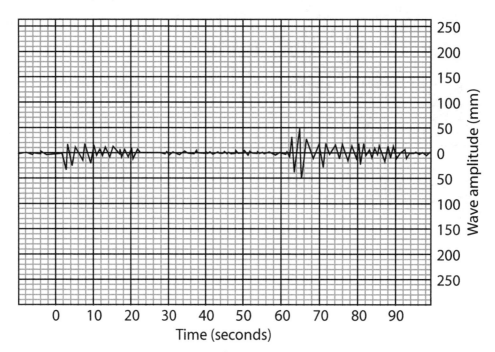

Figure 10.4

10-4

4. Locate the epicenter using the travel time curves.

a. On the map below (Figure 10.5) to draw the three circles with the radius corresponding to the epicentral distance.

b Which city was closest to the epicenter?

Figure 10.5

DETERMINATION OF THE RICHTER MAGNITUDE

The Richter Magnitude is a measure of earthquake strength based on the amplitude of S waves and the epicentral distance. A **nomogram** is a graphic device that estimates the magnitude based on the distance from the epicenter and the amplitude of the S waves measured from the seismogram (Figure 10.6). The epicentral distance is read from the travel-time curve on the corresponding S-P time delay. The amplitude is read from the seismogram as shown in Figure 10.7.

1. Calculate the magnitude. Once you have the epicentral distance and the S-wave amplitude, plot them on the nomogram and join the points with a line that crosses the magnitude bar. The value at which the line crosses is the magnitude for that earthquake. You can plot all three values and if your measurements were precise, you will obtain the same magnitude values.

The magnitude for the earthquake is:

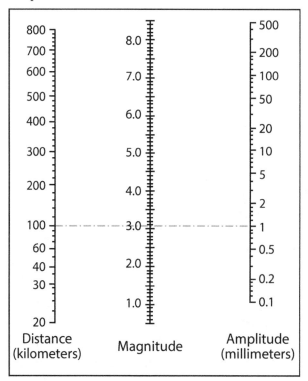

Figure 10.6

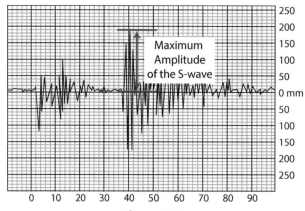

Figure 10.7

CASE STUDY: VIRGINIA SEISMICITY

The state of Virginia is located in a relatively stable portion of the north American tectonic plate. Virginia does not experience the amount and intensity of earthquakes that affect regions in close proximity of plate boundaries, like California. Historical records of earthquakes in Virginia go back to the eighteenth century, and though these earthquakes are not very strong, they are quite numerous. The North American tectonic plate, however, has some areas of lithospheric weakness that can accumulate geologic stress and release earthquakes.

Virginia Tech seismological observatory reports that Virginia has at least an earthquake a month though on average two per years are felt by the inhabitants near the epicentral area (http://www.magma.geos.vt.edu/vtso/va_quakes.html#history).

A note of caution when studying earthquake's strength. Magnitude was defined by C. Richter in 1935, so how do we know about the magnitude of historical earthquakes? If they occurred after the modern seismographs came into use (in the late nineteenth century), the seismograms of earthquakes can be compared with similar records for later earthquakes. For earthquakes that occurred before 1890, magnitudes are much harder to estimate. Intensity can be estimated by evaluating historical records of physical changes to the topography (i.e., faulting, mass wasting, river channel changes) and reports of damage and casualties and comparing them to the Mercalli modified scale or equivalent scales. There is not, however, a precise equivalence between intensity and magnitude, therefore magnitude of historical earthquakes can be estimated in a range of values (e.g., the Magnitude of the New Madrid [Missouri] earthquake main shock of the 1811–1812 seismic crisis was estimated between 6–8).

1. Examine the USGS Open File report 2006-1017 for the Seismicity of Virginia that can be downloaded here: https://pubs.usgs.gov/of/2006/1017/. Answer the following questions:

    a. Which is the oldest earthquake recorded in Virginia?

    b. What is the range of magnitude estimated for Virginia historical earthquakes?

    c. What are the three seismogenetic zones that Virginia belongs to? How were they identified?

2. Recent seismicity of Virginia. Go to: http://earthquake.usgs.gov/earthquakes/byregion/virginia.php and click on "all earthquakes, 1900 to present." The default will show you earthquakes since 1974. Make sure the legend (bottom center of the page) is visible and scan the table on the left side of the screen. Download the table (see tab to the top left for download options). Generate a spreadsheet that you can use to answer the following questions.

   a. When and where did the most recent earthquake occurred in Virginia? What magnitude? What depth?

   b. What is the range of magnitude of the earthquakes in your set?

   c. What is the range of depth?

   d. How does their location compare to the seismic zones you have described in answer to question 1.c?

3. On August 23, 2011, a magnitude 5.8 earthquake occurred in Louisa County, VA. The earthquake was widely felt. The seismograms were recorded by the network of instruments of the Virginia Tech seismological Observatory. You can see the seismograms recorded at three of the network stations here:

   http://www.magma.geos.vt.edu/vtso/2011/0823-louisa/seismograms1.png

   a. Which seismic station was the closest to the epicenter? How can you tell?

   b. Using the travel-time chart (Figure 10.1) calculate the distance from the epicenter for each station.

   RCRC: epicentral distance _____(km)

   VWCC: epicentral distance _____(km)

   BLA: epicentral distance _____(km)

   c. At what time did people in Blacksburg at VA felt the jolt of the P waves? The shaking of the S waves?

d. Examine the intensity map for this earthquake as compiled by the USGS based on the observations and damage report at: https://earthquake.usgs.gov/earthquakes/eventpage/se609212#shakemap

If you click on the map, a larger map with the primary source of info will appear. Compare the map with the Mercalli Modified Scale in this chapter (Table 10.1). How can you describe the effects in the area were the intensity was highest?

e. Does the map intensity correspond to the epicenter? Consider the distribution of population in the area

WORLD SEISMICITY

Use the following website to observe the most recent seismicity, http://ds.iris.edu/seismon/.
Click on the 30 days tab. A table will appear. Use that table and the map to answer the following questions

1. Distribution and magnitude of recent earthquakes
   a. Where did the ten largest earthquakes occur during the past month? Plot the epicenters in the map next page. Use different size symbols to indicate their magnitude, make a legend to explain your choice.

   b. Use different colors to differentiate the depth.

2. Sketch the plate boundaries on the map. How does the location of these earthquakes match position of the plate boundaries? Are all of your earthquakes located on plate boundaries?

3. Click on the Location tab for the strongest earthquake and you will see the distribution of the seismicity in that area by depth. Look at the right side of the screen, and click on the 3D menu item. At this point you will be able to see how the earthquake foci are distributed. What can you observe for this earthquake?

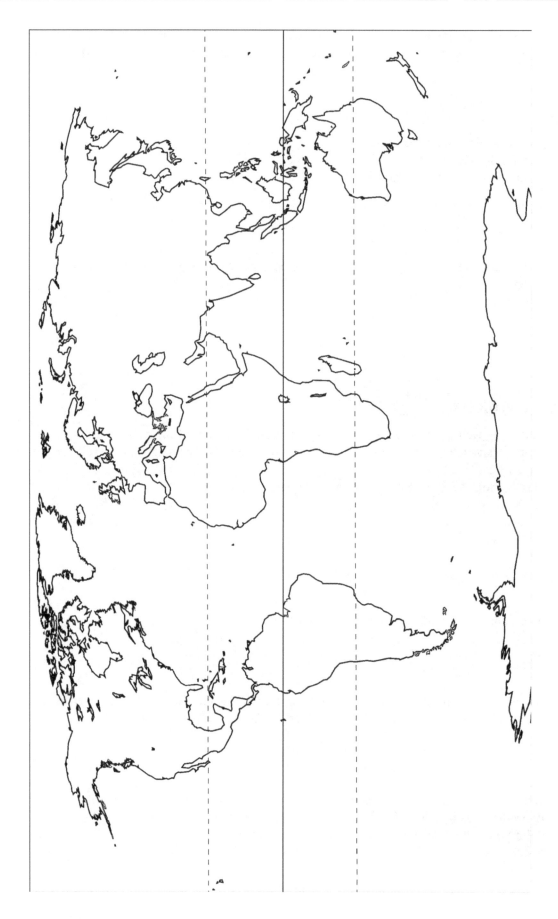

Student Name _____ Lab section _____ Date _____

CASE STUDIES: SEISMIC HAZARDS

Earthquake Intensity and Secondary Earthquake Hazards

The main hazard related to an earthquake is the ground shaking, however, earthquakes poses several other hazards that can have longer lasting and more devastating effects on the natural environment and human interests. These are called secondary seismic hazards:

- Mass wasting and soil liquefaction
  Earthquakes can trigger mass wasting, especially in areas with water-saturated soils. This includes soil liquefaction
- Damage to structures and infrastructures
  Buildings can collapse, trapping people inside and burying streets in rubble. Damaged roads, rails, bridges can disrupt or cut off entire communities and hinder rescue efforts
- Fires and loss of lifelines
  From damage to utilities lines
- Tsunami
  Greatest hazard at coastline communities.
- Permanent ground deformation
  Faults can permanently move the ground up or down
- Loss of life and societal disruption
  Casualties. Societal values and economy framework can be permanently affected.

The size of these secondary hazards depends on the earthquake strength and duration, on the local geological conditions and on the type of structures and society preparedness.

In the following exercise, you will evaluate the differences between measurements of intensity and magnitude and their relationship to secondary seismic hazards for case studies of historical earthquakes.
Read the material for each case study and answer the questions. For an in depth study click on the links to access the main source of info.
Based on the description, estimate the intensity of the earthquake main shake using the Modified Mercalli Intensity scale.

Table 10.1   Modified Mercalli Intensity Scale

Intensity	Shake	Description of events and damage
I	Not felt	Not felt except by a very few under especially favorable conditions.
II	Weak	Felt only by a few persons at rest, especially on upper floors of buildings.
III	Weak	Felt by persons indoors, especially on upper floors of buildings. Many people do not recognize it as an earthquake. Standing motor cars may rock slightly. Vibrations similar to the passing of a truck. Duration estimated.
IV	Light	Felt indoors by many, outdoors by few during the day. At night, some awakened. Dishes, windows, doors disturbed; walls make cracking sound. Sensation like heavy truck striking building. Standing motor cars rocked noticeably.
V	Moderate	Felt by nearly everyone; many awakened. Some dishes, windows broken. Unstable objects overturned. Pendulum clocks may stop.
VI	Strong	Felt by all, many frightened. Some heavy furniture moved; a few instances of fallen plaster. Damage slight.
VII	Very strong	Damage negligible in buildings of good design and construction; slight to moderate in well-built ordinary structures; considerable damage in poorly built or badly designed structures; some chimneys broken.
VIII	Severe	Damage slight in specially designed structures; considerable damage in ordinary substantial buildings with partial collapse. Damage great in poorly built structures. Fall of chimneys, factory stacks, columns, monuments, walls. Heavy furniture overturned.
IX	Violent	Damage considerable in specially designed structures; well-designed frame structures thrown out of plumb. Damage great in substantial buildings, with partial collapse. Buildings shifted off foundations.
X	Extreme	Some well-built wooden structures destroyed; most masonry and frame structures destroyed with foundations. Rails bent.
X		Few, if any, (masonry) structures remain standing. Bridges destroyed. Broad fissures in ground. Underground pipe lines completely out of service. Earth slumps and land slips in soft ground. Rails bent greatly.
XII		Damage total. Waves seen on ground surfaces. Lines of sight and level distorted. Objects thrown upward into the air.

CASE STUDY 1: December 16, 1811: New Madrid Fault estimated Richter value: 8.0

At the onset of the earthquake, the ground rose and fell—bending the trees until their branches intertwined and opening deep cracks in the ground. Landslides swept down the steeper bluffs and hillsides; large areas of land were uplifted; and still larger areas sank and were covered with water that emerged through fissures. Huge waves on the Mississippi River overwhelmed many boats and washed others high on the shore. High banks caved and collapsed into the river; sand bars and points of islands gave way; whole islands disappeared. Local uplifts of the ground and water waves moving upstream gave the illusion that the river was flowing upstream. Ponds of water also were agitated noticeably. Surface rupturing did not occur, however. The region most seriously affected was characterized by raised or sunken lands, fissures, sinks, sand blows, and large landslides that covered an area of 78,000–129,000 square kilometers, extending from Cairo, Illinois, to Memphis, Tennessee, and from Crowley's Ridge to Chickasaw Bluffs, Tennessee.

Although the motion during the first shock was violent at New Madrid, Missouri, it was not as heavy and destructive as that caused by two aftershocks about 6 hours later. Only one life was lost in falling buildings at New Madrid, but chimneys were toppled and log cabins were thrown down as far distant as Cincinnati, Ohio; St. Louis, Missouri; and in many places in Kentucky, Missouri, and Tennessee.

In Lake County uplift the Mississippi River valley was upwarped in several topography bulges. Other areas subsided by as much as 5 m, although 1.5–2.5 m was more common in Arkansas.

https://earthquake.usgs.gov/learn/topics/nmsz/1811-1812.php

    a. What was the intensity of the New Madrid Earthquake?

    b. List the evidence you based your answer on.

    c. Which of the items you described in your answer to b, is related to a secondary hazard?

CASE STUDY 2: April 18, 1906; San Francisco. Estimated Richter Magnitude: 7.8

The earthquake damaged buildings and structures in all parts of the city and county of San Francisco, although over much of the area, the damage was moderate in amount and character. Most chimneys toppled or were badly broken. The business district was built on ground filled in over Yerba Buena cove. Pavements were buckled, arched, and fissured; brick and frame houses were damaged extensively or destroyed; sewers and water mains were broken; and streetcar tracks were bent into wavelike forms.

On or near the San Andreas fault, buildings were destroyed and trees were knocked to the ground. The surface of the ground was torn and heaved into furrow-like ridges. Roads crossing the faultline were impassable, and pipelines were broken, shutting off the water supply to the city. The fires that ignited soon after the onset of the earthquake quickly raged through the city because of the lack of water to control them. They destroyed a large part of San Francisco.

Dislocation of fences and roads indicated the amount of ground movement between 3–4.5 m. In Mendocino County, a fence and a row of trees were displaced almost 5 m. Vertical displacement of as much as 0.9 m was observed in Sonoma County. Vertical displacement was not detected toward the south end of the fault.

Source: http://www.sfmuseum.org/1906/06.html

a. What was the intensity of the 1906 San Francisco earthquake?

b. List the evidence you based your answer on.

c. Which of the items you described in your answer to b, is related to a secondary hazard?

d. What are the strengths and weaknesses of the Modified Mercalli Scale when studying historical earthquakes?

CASE STUDY 3: Present day seismicity of the case study areas.

Ground shaking, the primary seismic hazard, is expressed as a probability of a given percentage of gravity acceleration. Examine the primary seismic hazard map for the United States (2014).
    https://pubs.er.usgs.gov/publication/sim3325
    Or download the pdf https://pubs.usgs.gov/sim/3325/pdf/SIM3325_sheet1.pdf

   a. Based on the values presented in this map, what is the peak ground acceleration (worst case scenario) level of seismic hazard for the San Francisco area? And for the New Madrid area?

   b. Compare and contrast the level and extent of the primary seismic hazard for both case studies. How are they related to the tectonic plates?

   Compare and contrast the seismicity in the two areas of the case studies.
   c. How frequent are earthquakes in the two areas considered for these case studies? Go to https://earthquake.usgs.gov/earthquakes/map/
   Check the settings to display the earthquake for the past 30 days in the United States.

Laboratory Experience Assessment: Earthquakes Analysis

Learning Objective	Level of Confidence		
	Hesitant, concept is unclear, would not know how to use/apply	I have a general idea of what this is about and with guidance I could apply what I learned to problem solving	I am confident I understand this topic and I can apply it to solving a problem
Identify P and S waves of a simple seismogram			
Locate the epicenter of an earthquake using the seismograms and travel-time curves			
Explain the basis for determining the Richter magnitude and determine it using the Normogram			
Find, describe and elaborate on recent seismic events from data sets available through USGS			
How challenging did you find this exercise?			
What challenged/interested you the most about this activity? Why?			

Chapter 11

Crustal Deformation

PRELAB STUDY SESSION

In preparation to this exercise, briefly answer the following questions. Refer to your textbook for the information you need. In addition to these questions, your lab instructor might give you more questions

A. Sketch an anticlinal fold. Be sure to note the limbs of the fold and the fold axial plane on your sketch.

B. Does the above fold form from compression or extension?

C. Sketch a normal fault. Note the hanging wall and the footwall.

D. Does the above fault represent a compressional or extensional environment?

E. Sketch a Strike and Dip symbol representing N15E. (North is at the top of the page.)

F. Now sketch a Strike and Dip symbol representing W45S.

ORIENTATION OF STRUCTURES

A description of the *attitude* (orientation in space) of geologic structures is given in terms of strike and dip (Figure 11.1). Strike represents the direction of a geologic surface with respect to the north and is measured on a horizontal surface, while dip is the angle between the measured planes.

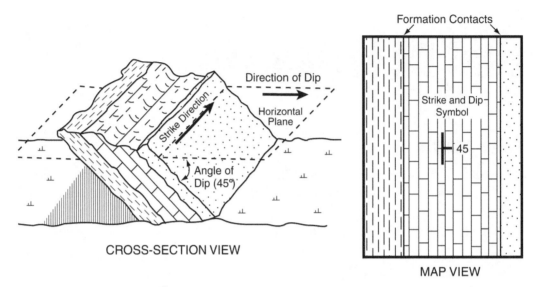

Figure 11.1   Measurement of strike and dip.

FOLDS

Folds are geologic structures generated by compression. The fold axis is the line marking the position of the sharpest bend of the fold, also known as the hinge line. The axial plane divides the fold at its maximum curvature (Figure 11.2).

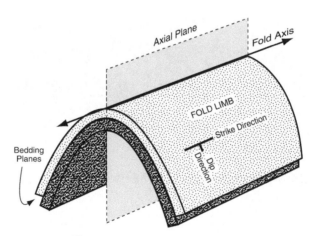

Figure 11.2   Fold nomenclature.

There are two fundamental classes of folds: the *anticline*, or up-turning fold, and the *syncline*, or down-turning fold. In anticlines, the youngest rocks are along the edges of the fold, and in synclines the youngest rocks are in the middle of the fold (Figure 11.3).

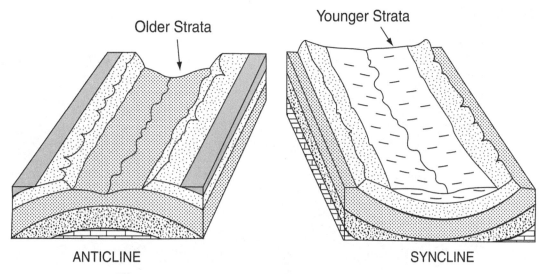

Figure 11.3 Block diagrams of an anticline and a syncline.

Folds may be **symmetrical**, with limbs dipping at equal angles in opposite directions: **asymmetrical**, with limbs dipping in opposite directions at unequal angles; **overturned**, with both limbs dipping in the same direction; or **recumbent**, with the axial plane nearly horizontal (Figure 11.4).

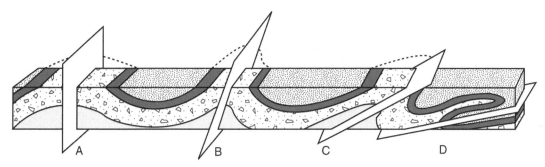

Figure 11.4 Block diagram in which four variations of a fold are shown: A. Symmetrical anticline; B. Asymmetrical anticline; C. Overturned anticline; D. Recumbent fold. Note the different attitudes of the four axial planes.

The flanks of the fold are called limbs. When the hinge line is not horizontal, the fold plunges (Figure 11.5).

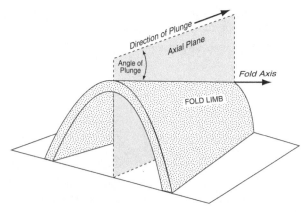

Figure 11.5 Plunging anticline fold.

OUTCROP PATTERNS

Horizontal strata: Undeformed horizontal rocks exhibit characteristic map patterns that parallel topographic contour lines. Gentle slopes have wider outcrop patterns, and steep slopes have narrower outcrop patterns (Figure 11.6).

Perspective View

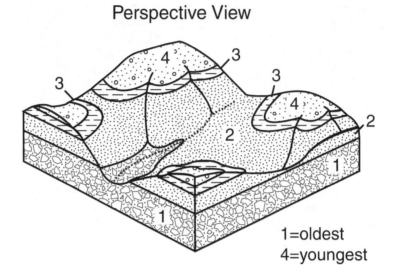

1=oldest
4=youngest

Map View

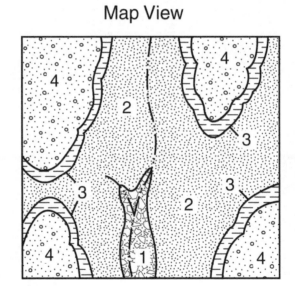

Figure 11.6 Perspective and map view of horizontal strata.

Inclined strata: Tilted layers of rock form rough parallel bands in map view. The width of the outcrop patterns depends on the dip of the strata. Gentle dipping beds have wider outcrop patterns than steeply dipping beds over the same thickness. When intersected by a stream, the dipping beds form a "V" in the direction of the dip (Figure 11.7).

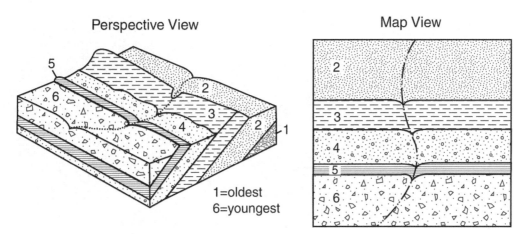

Figure 11.7 Perspective and map view of inclined strata.

Folded strata: Strata are parallel in map view. Younger rocks are exposed in the center of the synclines, while older rocks are exposed in the center of anticlines (Figure 11.3).

Plunging folds have a characteristic "V" shaped outcrop on a geologic map. The plunge of the fold axis is drawn with an arrow indicating the plunge direction. On geologic maps, a bold line represents the axis of the fold, while smaller arrows perpendicular to the axial line, indicate the direction in which the limbs dip. Anticlines plunge toward the narrowest part of the fold, while synclines plunge toward the widest part of the fold (Figure 11.8).

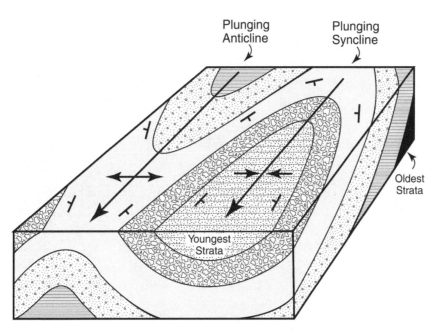

Figure 11.8 Outcrop patterns of plunging folds with strike and dip symbols, fold axis, and plunge directions added to the map view.

Dome: An area of strata with a point of maximum curvature. On a geologic map, a dome looks like circular outcrop pattern with the oldest rocks exposed in the center (Figure 11.9).

Block Diagram
Perspective View

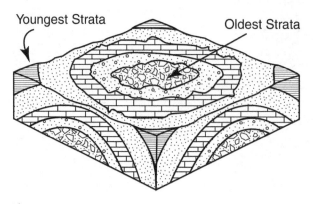

Figure 11.9   Perspective and map view of domed strata.

Basins: Structural basins form outcrop patterns that look similar to domes but the youngest rocks are exposed in the center. Stream cuts form a "V" toward the center of the structure (Figure 11.10).

Block Diagram
Perspective View

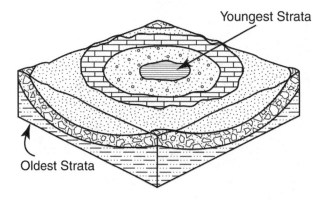

Figure 11.10   Perspective and map view of a structural basin.

FAULTS

Faults are fractures in the crust where displacement as occurred. These structures can form from compression, extension, or shear stress.

Normal faults: Form when vertical displacement occurs from extension by tensional stress. These primarily occur at rifting regions. One block, the hanging wall, moves down the surface of the fault relative to the foot wall. These faults lengthen and thin the crust (Figure 11.11).

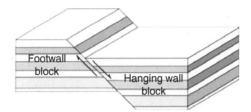

Figure 11.11   Normal fault

Reverse faults: Form from vertical displacement generated by compression. The hanging wall moves up relative to the footwall. In these faults, the crust is thickened and shortened. A reverse fault at a low angle and extensive displacement is a thrust fault and is always associated with mountain building (Figure 11.12).

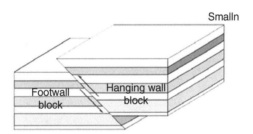

Figure 11.12   Reverse fault

Strike-slip faults: These faults are produced by shear stress, the resulting movement is a side-to-side displacement of the blocks (Figure 11.13).

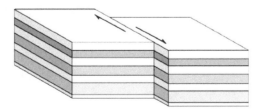

Figure 11.13   Right lateral strike-slip fault

On geologic maps, faults are represented by heavy lines, and an offset in strata

GEOLOGIC MAP SYMBOLS

Geologic maps show the distribution of different types of rocks and geologic structures' folds and faults. A geologic map is printed on top of a topographic map. Different rocks are represented by coded colors, and geologic structures are represented by lines and symbols shown in Figure 11.14.

Google Map Symbols

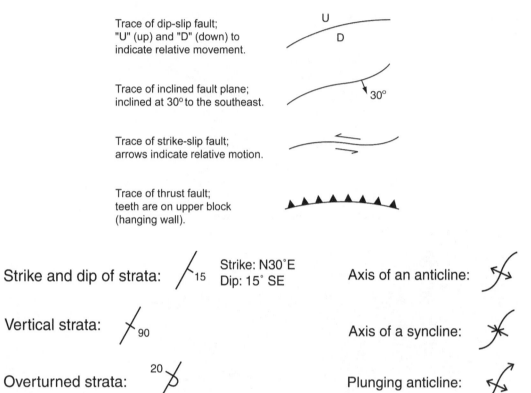

Figure 11.14   Common geologic map symbols

Student Name _____ Lab section _____ Date _____

LABORATORY EXERCISE

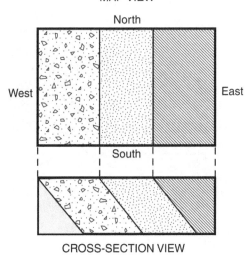

MAP VIEW

North

West

East

South

CROSS-SECTION VIEW

PERSPECTIVE VIEW

North

West

Diagram 11A.

1. **Refer to Diagram 11A for the following exercise:**

   a. What is the direction of strike? _____; of dip? _____

   b. Place the proper strike and dip symbol on the *map view.*

   c. Place the proper strike and dip symbol on the *perspective view.*

   d. What is the angle of dip? _____

   e. On which view can the dip angle be accurately measured (map view, perspective view, or cross section)? _____

2. **Add correctly oriented strike and dip symbols to Diagram 11B:**

(a)

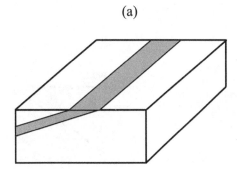

(b)

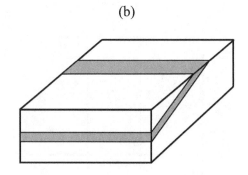

(c)

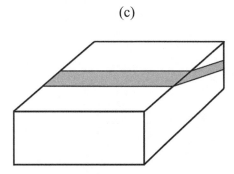

(d)

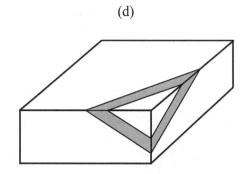

(e)

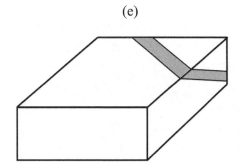

(f)

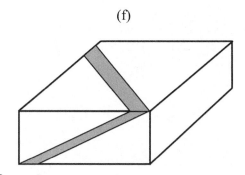

Diagram 11B.

3. **Refer to Diagram 11C for the following exercise:**

   a. Fill in the map view based on the layer orientation in the cross section.

   b. The layer has the same thickness in each cross section; the only difference is the dip angle. How does the dip angle affect the map-view expression of the layer?

   c. At what dip angle does the map expression indicate the true thickness of the layer?

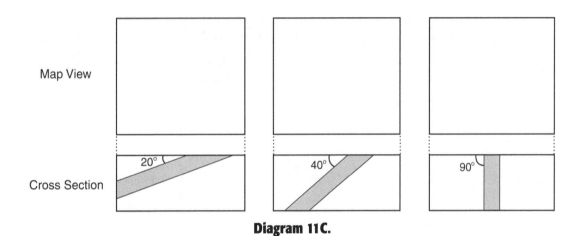

Diagram 11C.

4. **Refer to Diagram 11D for the following exercise:**

   a. What is the map direction of strike? _____

   b. Which direction do the beds dip on the east side of the map? _____

   c. On the west side? _____

   d. Draw the cross-section in the space provided.

   e. What is the structure? _____ (anticline or syncline)

   f. Which rocks are the oldest? _____ (shale, sandstone, or conglomerate)

   g. Do the dip symbols point toward the oldest or youngest beds? _____

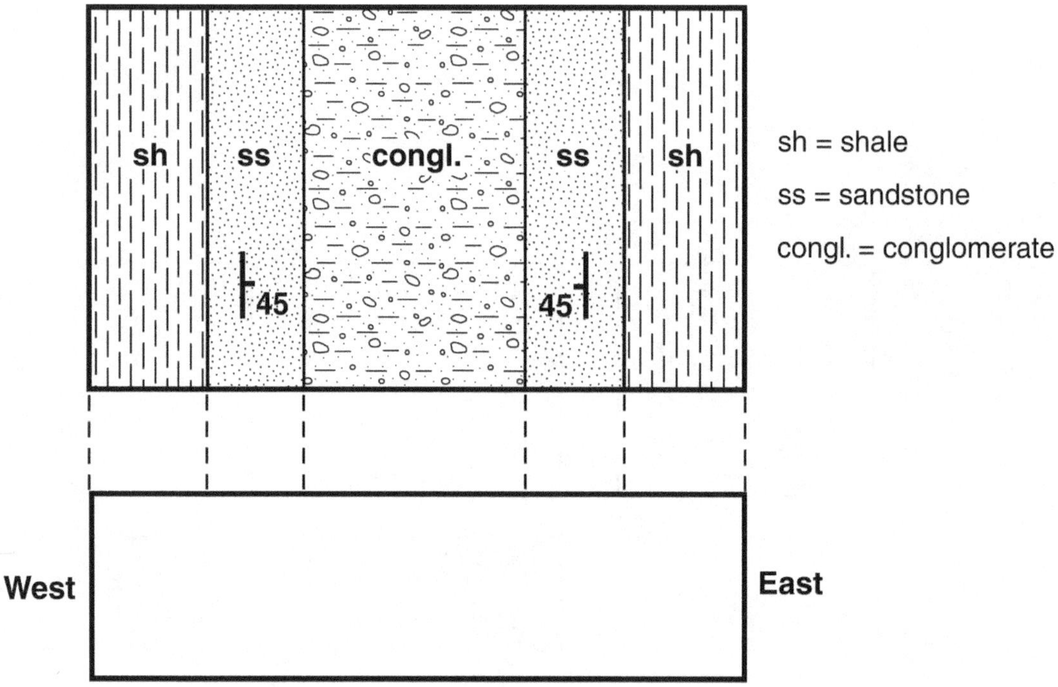

sh = shale

ss = sandstone

congl. = conglomerate

West   East

Diagram 11D.

5. **Refer to Diagram 11E for the following exercise:**

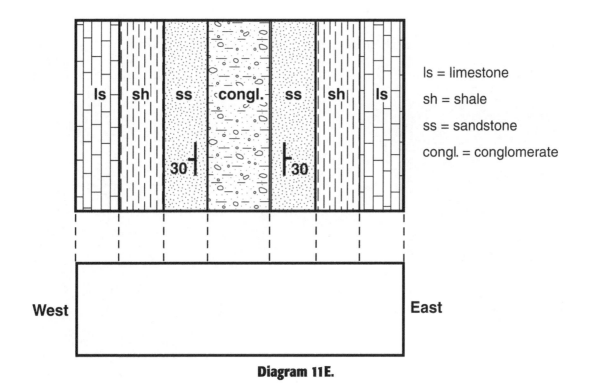

ls = limestone

sh = shale

ss = sandstone

congl. = conglomerate

West   East

Diagram 11E.

a. Draw the cross-section. What is the structure? _____

b. Is the structure plunging? _____

c. How do you know? _____

d. On the cross-section, number the beds in order: 1, 2, 3, 4; 1 for the oldest, 4 for the youngest. Also number the beds on the map view.

e. Do the dip symbols point toward the oldest or the youngest beds? _____

6. **Refer to Diagram 11F for the following exercise:**

   a. Complete the map view. Include proper bed numbers.

   b. Show the map symbols to the completed diagram.

   c. What is the *direction* of the plunge? _____

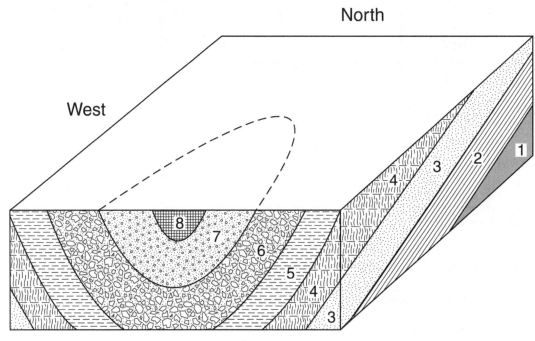

Diagram 11F.

7. **Refer to the figures in Diagram 11G for the following exercise:**

The numbers on each map indicate the relative ages; 1 being the oldest.

a. Complete the cross-sectional views and give the name of the geologic structure(s) present.

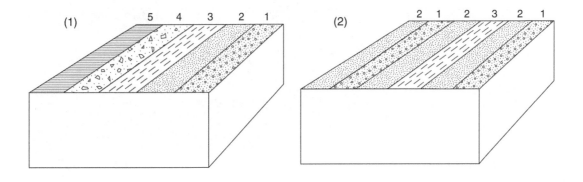

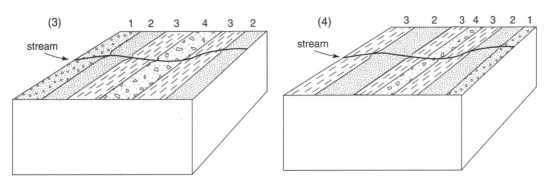

Diagram 11G.

8. In Diagram 11H below, label the following:

   a. Hanging wall block
   b. Footwall block
   c. Strike of the fault plane
   d. Displacement
   e. Arrows indicating *relative* movement of fault blocks
   f. What type of fault is illustrated?

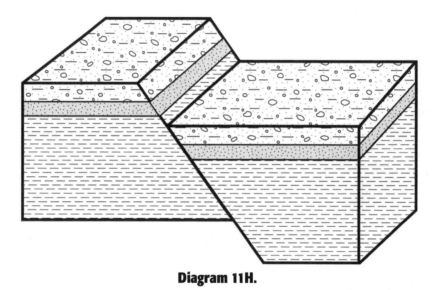

Diagram 11H.

9. In Diagram 11I below, label the following:

    a. Hanging wall
    b. Foot wall
    c. Strike of the fault plane
    d. Displacement arrows indicating relative movement of the fault blocks
    e. What type of fault is illustrated in this diagram?

    f. What would happen if the sandstone of the up-thrown block eroded down to the level of the down-thrown block?

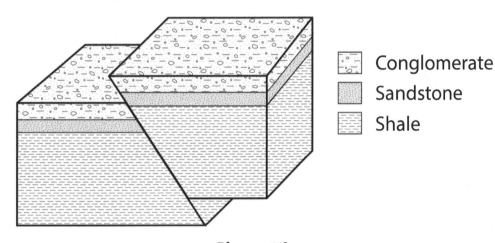

        ⊡ Conglomerate
        ▨ Sandstone
        ▤ Shale

Diagram 11I.

10. Effects of erosion on faulted rocks

Examine the block diagrams in 11J. The left column shows the blocks before faulting; the dashed line indicates the position of the fault trace. The diagrams in the middle column show the relationship immediately after faulting.

a. Complete the third blocks (top and sides) in the column on the right to illustrate the surface after the up-thrown part of the blocks have been eroded down to the same level as the down-thrown block.

b. Label the elements of each block and show the arrows indicating the relative movement of each block.

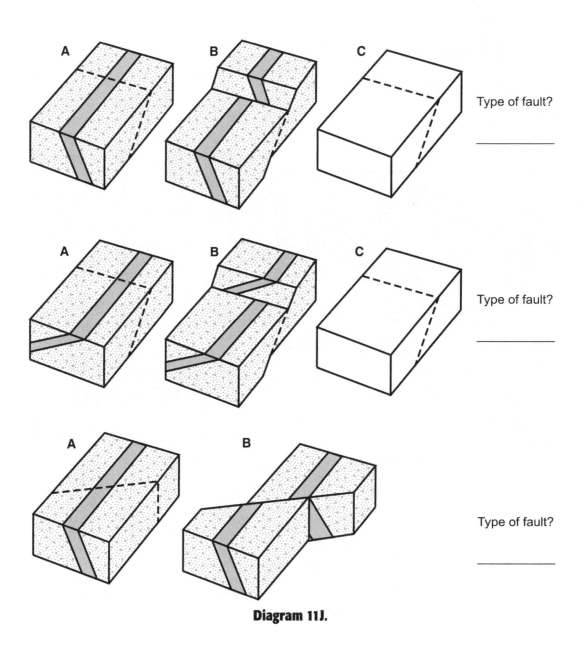

Diagram 11J.

11. Sketch in the pattern on the ***upthrown*** block in the Diagrams 11K. (Also sketch in the fold axis.)

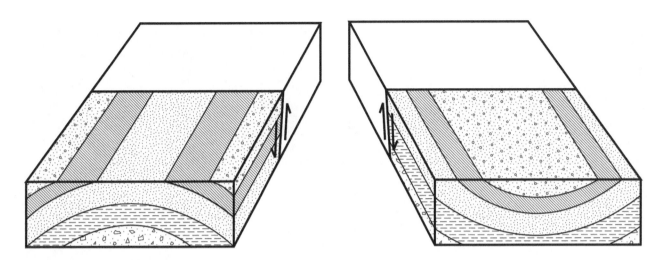

Diagram 11K.

12. Assemble the block models A-E on the following pages. NOTE: Layer #1 represents the oldest layer on each box.
    a. Cut along the outline and fold the side tabs and/or the bold lines.
    b. When the boxes are complete, fill in the blank sides.
    c. For Box Models A–D: Add the strike and dip, and the pertinent map symbols (described in this chapter) on the TOP of each box model.
    d. For Box Models E: Sketch sedimentary strata on the SIDES of both models to align them. These will illustrate TWO different types of displacement.
    e. When you have completed the boxes, be sure to note the correct geologic structure you have just created on the top of the boxes.

BOX MODEL A

1 2   3 5

4

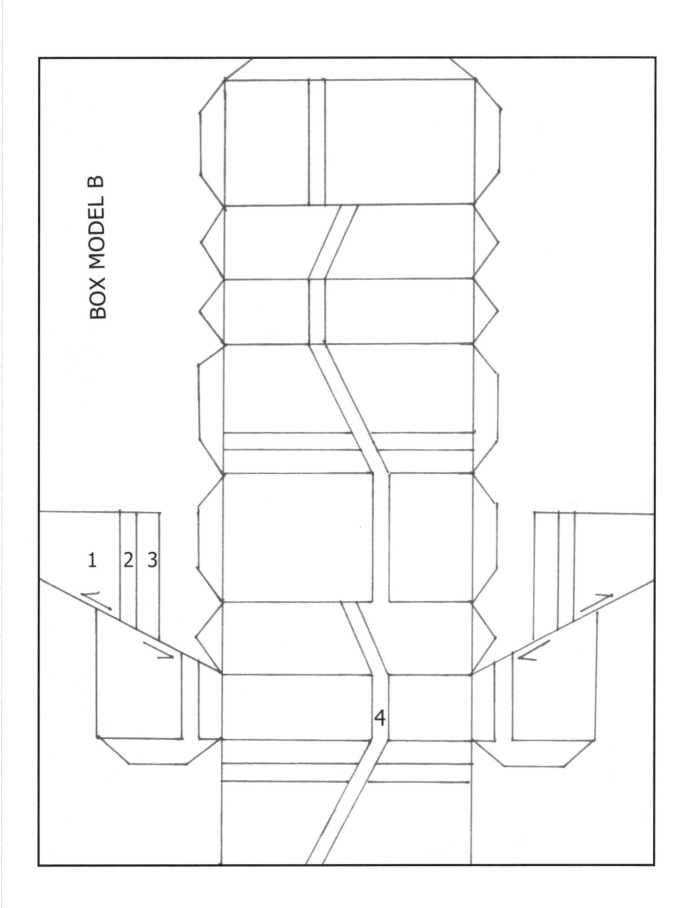

BOX MODEL B

1 2 3

4

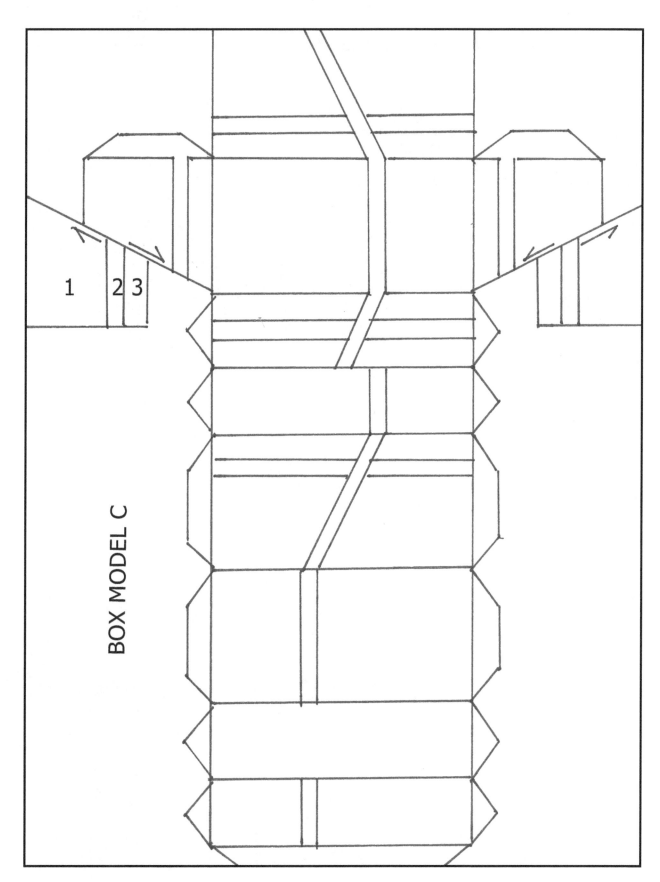

BOX MODEL C

1 2 3

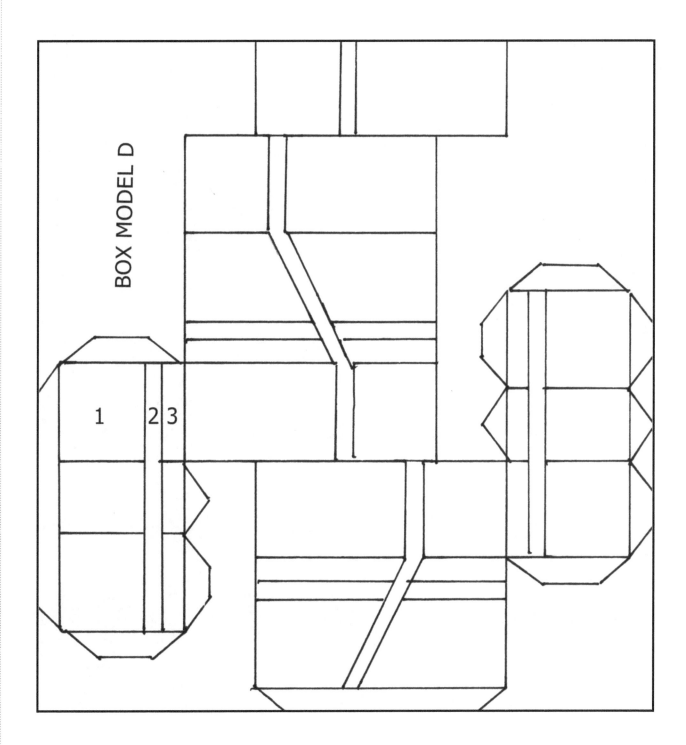

BOX MODEL D

1 2 3

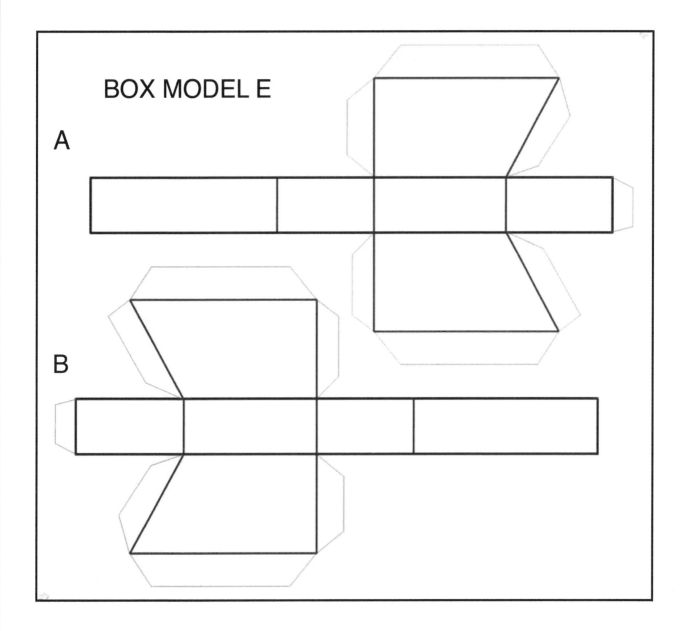

BOX MODEL E

A

B

Laboratory Experience Assessment: Geologic Structures

Learning Objective	Level of Confidence		
	Hesitant, concept is unclear, would not know how to use/apply	I have a general idea of what this is about and with guidance I could apply what I learned to problem solving	I am confident I understand this topic and I can apply it to solving a problem
Recognized structures formed by compression, tension and shear			
Define, sketch and recognize on a block diagram the basic types of folds and faults			
Define and use strike and dip including reporting these on a simplified diagram and/or map			
On a simplified geologic map, recognize the basic types of folds and faults based strike and dip and relative position of rock strata			
How challenging did you find this exercise?			
What challenged/interested you the most about this activity? Why?			

Chapter 12

Plate Dynamics and Isostasy

PRELAB STUDY SESSION

In preparation to this exercise, briefly answer the following questions. Refer to your textbook for the information you need. In addition to these questions, your lab instructor might give you more questions to answer.

A. Explain the process of isostasy, in relation to continental and oceanic crust. Is this significantly different when relating it to glaciers? Explain.

B. What is the source of tectonic plate movement?

C. What do Yellowstone National Park, Galapagos Island, and the Hawaiian Islands have in common? Explain how these features form.

D. List the seven large tectonic plates that are found on the Earth today.

E. What type of plate boundary is closest to George Mason University? Where is it?

PLATE MOVEMENT CASE STUDY: HAWAIIAN HOT SPOT

Hot spots were initially explained as the result of the interaction between the lithosphere and deep-rooted thermal anomalies in the deep mantle. These thermal anomalies, or mantle plumes, were believed to rise with steady and significant heat, and until the end of the 20th century, hot spots were thought to be very stable features, active for hundreds of million years. More recent geophysical evidence indicates that they are not quite as deep or as permanent, but they are still deep rooted in the mantle and more permanent than tectonic plates, so when the tectonic plates move above mantle plumes, hot spots form.

By visual examination of the Pacific Ocean sea floor, we can see that the Hawaiian Islands are aligned in a northwest trending direction. Beyond the islands, the trend continues with a sequence of aligned underwater extinct volcanoes (called seamounts) that extend for hundreds of kilometers. A third of the way across the Pacific sea floor, the line of predominantly underwater volcanoes continues, but the alignment changes direction, pointing towards the north. This alignment of volcanoes is called the Emperor Seamount chain (Figure 12.1).

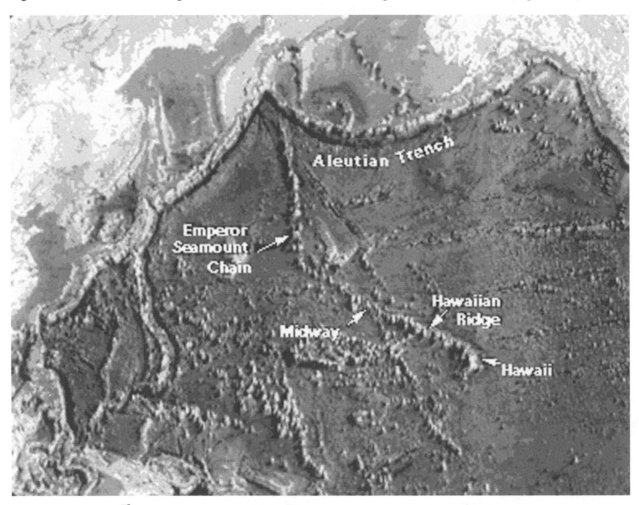

Figure 12.1 Location of Hawaiian-Emperor volcanoes in the Pacific Ocean.

In addition to the alignment of seamounts, other evidence comes from the ages of the volcanic rocks, which become increasingly older as they get further away from the present day active Hawaiian volcano (Kilauea), as well as from the chemistry of the rocks that, though not identical, have compositions that belong to the same type of magmas.

Table 12.1 lists the distance from Kilauea and the age of several volcanic islands along the seamount chain. The error ranges noted near the ages of the volcano depends on the quality of the sample from the seamount. Because these are underwater volcanic structures some samples are difficult to acquire, and at times the only available samples are altered; therefore, the error on the age determination is larger.

Table 12.1 Distance from the Hawaiian Island of Kilauea and radiometric age of volcanic islands along the Hawaiian-Emperor seamount chain.

Volcano Name	Distance from Kilauea along Trend of Chain (km)	Radiometric K-Ar Age in Million of Years (with error)
Kilauea	0	0-0.4
Mauna Kea	54	0.375 + 0.05
Kohala	100	0.43 + 0.02
Haleakala	182	0.75 + 0.04
Kahoolawe	185	> 1.03 + 0.18
West Maui	221	1.32 + 0.04
Lanai	226	1.28 + 0.04
East Molokai	256	1.76 + 0.04
West Molokai	280	1.90 + 0.06
Koolau	339	2.6 + 0.1
Waianae	374	3.7 + 0.1
Kauai	519	5.1 + 0.20
Niihau	565	4.89 + 0.11
Kaula	600	4.0 + 0.2
Nihoa	780	7.2 + 0.3
Unnamed seamount 1	913	9.2 + 0.8
Unnamed seamount 2	930	9.6 + 0.8
Necker	1.058	10.3 + 0.4
La Perouse Pinnacles	1.209	12.0 + 0.4
Brooks Bank	1.256	13.0 + 0.6
Unnamed seamount 3	1.330	13.0 + 0.6
Gardner Pinnacles	1.435	12.3 + 1.0
Unnamed seamount 4	1.460	12.3 + 1.0
Laysan	1.818	19.9 + 0.3
Northampton Bank	1.841	26.6 + 2.7
Pearl and Hermes Reef	2.281	20.6 + 2.7
Midway	2.432	27.7 + 0.6
Unnamed seamount 5	2.600	28.0 + 0.4
Unnamed seamount 6	2.825	27.4 + 0.5
Colohan	3.128	38.6 + 0.3
Abbott	3.280	38.7 + 0.9
Daikakuji	3.493	42.4 + 2.3
Yuryaku	3.520	43.4 + 1.6
Kimmei	3.668	39.9 + 1.2
Koko (southern)	3.758	48.1 + 0.8
Ojin	4.102	55.2 + 0.7
Jingu	4.175	55.4 + 0.9
Nintoku	4.452	56.2 + 0.6
Suiko (southern)	4.794	59.6 + 0.6
Suiko (central)	4.860	64.7 + 1.1

Pacific plate rate and direction of motion from seamounts data

Use graph paper or an excel spreadsheet to carry out this exercise.

Plot the distance from Kilauea Island versus the age of the volcanism for the islands in Table 12.1. Use the slope of the best-fit line to answer the following questions:

a. What has been the average rate of movement of the Pacific Plate for the last 70 million years?

b. Was the rate of movement uniform? Report the fastest and the slowest rate of movement. When did they occur? Which Islands were formed during these episodes?

c. What does the "bend" in the Hawaiian-Emperor chain indicate about the direction of movement of the Pacific Plate?

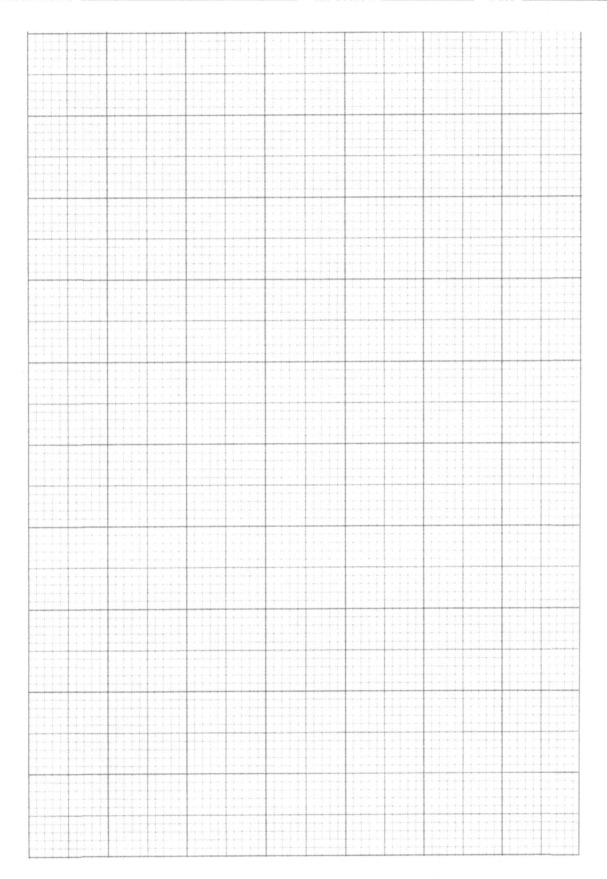

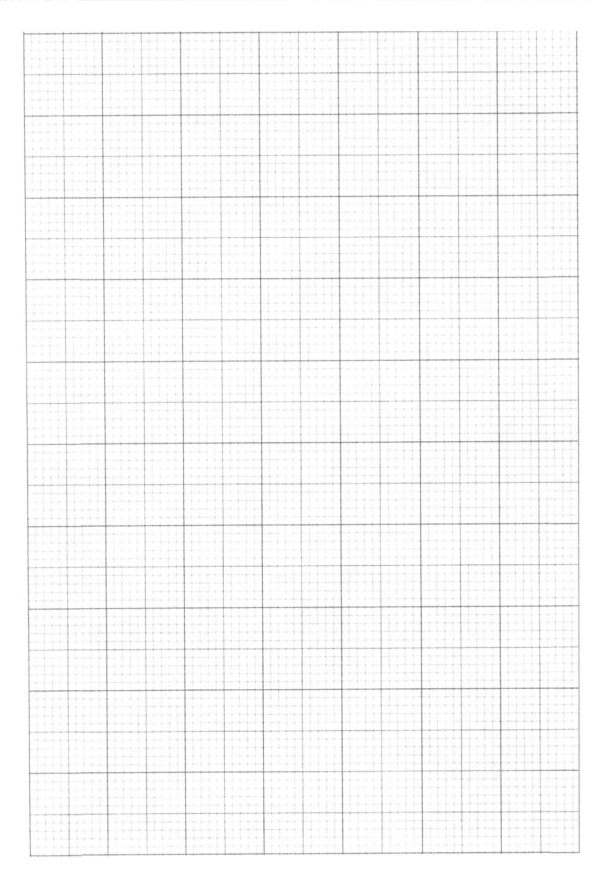

WORLD HOT SPOT CASE STUDY: EARTHREF DATABASE

EarthREF database has a census with the largest collection of seamounts. Visit the page at: http://earthref.org/ SC/. Seamount and seamount chains are highlighted. In order to find the info for each seamount, you need to zoom in close to the volcano.

1. Zoom to the Kilauea region: describe the distribution of smaller seamounts near the Hawaiian island.

2. Describe the general trends of seamounts in the Pacific. How do they compare to the Hawaiian-Emperor chain? In which way are seamount chains similar? In which way are they different?

3. The Samoan Islands are part of Polynesia. The islands are comprised of a chain of seamounts. Plot the data from Table 12.2 as you did for the data of Table 12.1. Distance is set from Vailulu'u seamount, the most active Samoan submarine volcano.

Table 12.2 Distance and age of samoan volcanoes

Seamounts	Distance from Vailulu'u	Age (MY)
Vailulu'u	0	0
Ta'u	20	0.3
Malu Malu	78	1.1
Ofu	79	1.1
Muli	114	1.6
Tulaga	115	1.6
Soso	136	1.9
Tutuila	153	2.2
Tamai'i	168	2.4
Tisa	247	3.5
Savai'i	416	5.9
Pasco	632	9.0
Savai'i Wallis Island	815	11.6
Combe	906	12.9

4. Is it possible to come to the same conclusions as for Hawaii? Why or why not?

5. Go to http://earthref.org/SC/ http://pubs.usgs.gov/gip/dynamic/world_map.html.

   Observe the distribution of hot spots on Earth. Can you find any other hot-spot-related feature that helps you track the direction of plate motion? Explain.

ISOSTASY CASE STUDY: ANTARCTICA

Isostasy is the gravity-controlled equilibrium between the lithosphere and the asthenosphere. Because of isostasy, lithospheric plates "float" above the asthenosphere. The "floating" elevation depends on the thickness and density of the lithosphere and its overburden, regardless if these are sediments or ice.

　　Examine Figure 12.2 and answer the following questions about the effect of Antarctic continental glaciers on the isostatic equilibrium of the lithosphere.

1. What is the maximum thickness of ice in Antarctica, and where is it found?

2. What is the average thickness of Antarctica ice?

3. Ice has a density of 0.9 g/cc. What is the mass of a 1 cm^3 of ice?

4. If a continental glacier melts, its weight is lifted from the lithosphere, which then "rebounds" upward. Mantle material will flow below the lithosphere to compensate for the loss of mass from the melting ice, but because mantle is denser than ice, the volume of mantle material needed to compensate for the loss of mass will be significantly lower. Considering that the average upper mantle rock is 3.3 cm^3, how much rebound can we expect from the crust if the ice melts?

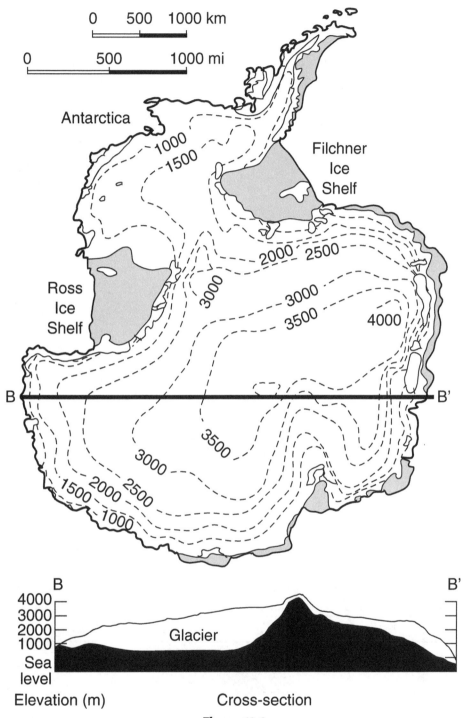

Figure 12.2

5. Compare oceanic and continental lithosphere in Figure 12.3. Column 1 represents an idealized section across the continental lithosphere. Column 2 represents an idealized section across the oceanic lithosphere composed, from the surface down, by seawater, sediments, basalt, and gabbro. Complete the tables 12.3 a and b below.

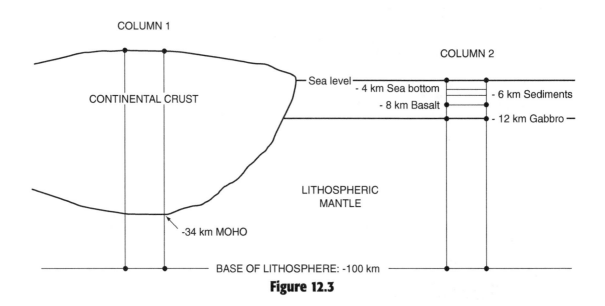

Figure 12.3

Table 12.3a
Continental lithosphere—Column 1

Continent	Density g/cc	Thickness	Mass
Crust	2.8		
Mantle	3.3		
Total Mass			

Table 12.3b
Oceanic Lithosphere—Column 2

Continent	Density g/cc	Thickness	Mass
Ocean Water	1.0		
Sediment	2.0		
Basalt	2.8		
Gabbro	3.0		
Mantle	3.3		
Total Mass			

6. How does the mass in columns 1 and 2 compare?

7. Think about geologic processes you know. How can the mass in the columns increase?

8. How can the mass in each column be decreased?

9. How do your calculations support the hypothesis of isostatic equilibrium? Explain what your calculations mean.

Laboratory Experience Assessment: Plate Dynamics and Isostacy

Learning Objective	Level of Confidence		
	Hesitant, concept is unclear, would not know how to use/apply	I have a general idea of what this is about and with guidance I could apply what I learned to problem solving	I am confident I understand this topic and I can apply it to solving a problem
Calculate plate motion rates and direction of movement based on evidence of movement			
Explain hot-spot volcanism evolution at the Hawaiian hot spot			
Collect evidence from similar processes (hot spot activity worldwide) and explain the process in general terms.			
Evaluate from data and measurements the effect of ice caps on buoyancy (Antarctica example)			
Explain how gravity, mass and buoyancy relate to plate tectonics			
How challenging did you find this exercise?			
What challenged/interested you the most about this activity? Why?			

Chapter 13

A Visit to the National Museum of Natural History

This self-guided visit starts at Hope Diamond in the Harry Winston Gallery. This is on the second floor of the Museum.

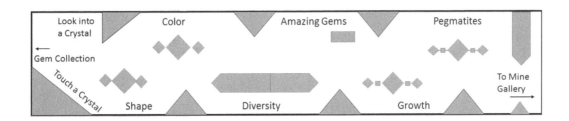

1. There are five large specimens along the wall. Describe which of the Earth Treasures is your favorite. Take a picture. Where was it found?

2. How did the Hope diamond get its name?
   a. How and where was it found?

   b. How old it is?

   c. What chemical element gives the Hope Diamond its color?

    d. What is a carat?

    e. How many carats does it weigh?

NATIONAL GEM COLLECTION HALL

3. Examine the center display case with large Topaz. What are the large bubbles in the crystals?

4. What is the relationship between Ruby and Sapphire?

DIAMONDS

5. Take a few minutes to examine the jewels showcases in this room. Describe your favorite jewel. Take a picture. What is the main gemstone? How many carats does it weigh? and it is set with other gems? What are they?

Now go into the Eberly Mineral and Gem Gallery

6. Look at the blue and red mural. What does it represent?

Next follow the wall of minerals to SHAPE: the many faces of crystals

7. Why are there such a large variety of crystal shapes?

8. Find a specimen in the "shape" section that you like. Take a picture and describe it. Where is it from? What is its chemical composition?

Now go to COLOR: Mineral Rainbow section.

9. Find a specimen in the "color" section that you like. Take a picture and describe it. Where is it from? What is its chemical composition?

Now go to GROWTH:

10. Why are some of these crystals so large and well-formed?

Turn around to center growth display.

11. What is the significance of Smithsonite?

12. What is an inclusion? Name one you see in the exhibit and take a picture.

13. Look at the GYPSUM giants. How did they get so big?

Next go to PEGMITITES: Big Crystals.

14. Choose a mineral in the Pegmitite that is not a common silicate. Describe it and take a picture.

15. Tourmalines are piezoelectric. What does that mean?

16. What is a pocket pegmatite? Watch the video and explain.

Go to the right to THE MOREFIELD MINE

17. a. In which type of rock is the mine extracting ore?

    b. What is Amazonite?

18. BISBEE MINE has produced about 200 types of minerals. Which are the most economically important?

19. What is so special about the SHIRLEY HILL MINE? Describe and take a picture of your favorite fluorescent mineral.

Enter the PLATE TECTONICS Hall. Examine the displays on your left.

20. How does Hot Spot activity differ between Hawaii and Yellowstone?

21. Now turn around to the volcanic activity at Divergent Plate Boundaries-Spreading Ridges.
    a. What specimens in this display come from the East Pacific Rise?

    b. How did it form?

Proceed forward to the beyond the big globe to "Where plates come together"

22. a. Look at the "water in water out" panel. What does this refer to? How does water affect volcanic eruptions?

    b. How much water is contained in the Andesite in front of you?

23. Continue to the earthquake study station. Watch the seismographs and the video display that is refreshed every 10 minutes.
    a. Mark the date and time of your visit.

    b. Look at the display screen, have there been any earthquakes in the United States today? Where?

Now move to your right, find the large granite block with bull's-eye targets. Tap the targets on the granite block.

24. a. Describe what shows up on the display when you hit the granite at various   strengths.

    b. Step a few feet away from the block and jump! Repeat by moving further away. How does the signal change?

You have completed your assignment. Now you can enjoy the rest of the Museum on your own.